CONFRONTING GLOBAL WARMING

I Water and Ice

CONFRONTING GLOBAL WARMING

Water and Ice

Noah Berlatsky

Michael E. Mann
Consulting Editor

GREENHAVEN PRESS
A part of Gale, Cengage Learning

Detroit • New York • San Francisco • New Haven, Conn • Waterville, Maine • London

Christine Nasso, *Publisher*
Elizabeth Des Chenes, *Managing Editor*

© 2011 Greenhaven Press, a part of Gale, Cengage Learning

For more information, contact:
Greenhaven Press
27500 Drake Rd.
Farmington Hills, MI 48331-3535
Or you can visit our Internet site at gale.cengage.com.

ALL RIGHTS RESERVED
No part of this work covered by the copyright herein may be reproduced, transmitted, stored, or used in any form or by any means graphic, electronic, or mechanical, including but not limited to photocopying, recording, scanning, digitizing, taping, Web distribution, information networks, or information storage retrieval systems, except as permitted under Section 107 or 108 of the 1976 United States Copyright Act, without the prior written permission of the publisher.

For product information and technology assistance, contact us at
Gale Customer Support, 1-800-877-4253.

For permission to use material from this text or product, submit all requests online at
www.cengage.com/permissions.

Further permissions questions can be emailed to permissionrequest@cengage.com

Every effort is made to ensure that Greenhaven Press accurately reflects the original intent of the authors. Every effort has been made to trace the owners of copyrighted material.

Cover Image copyright JinYoung Lee, 2010. Used under license from Shutterstock.com and © 2010 Photos.com, a division of Getty Images; Leaf icon © iStockPhoto.com/domin_domin.

LIBRARY OF CONGRESS CATALOGING-IN-PUBLICATION DATA

Berlatsky, Noah.
  Water and ice / Noah Berlatsky.
     p. cm. -- (Confronting global warming)
  Includes bibliographical references and index.
  ISBN 978-0-7377-4861-1 (hardcover)
  1. Sea level. 2. Ice. 3. Glaciers. 4. Global warming. I. Title.
  GC89.W38 2010
  551.45'8--dc22   2010011348

Printed in the United States of America
2 3 4 5 6 7 14 13 12 11

Contents

Preface   1
Foreword   3

Chapter 1: Water and Climate: An Introduction   10

The Hydrologic Cycle   10
    Table: Location and Form of Water on Earth   12
The Ocean Affects Climate   13
Water, Climate, and People   14
    Sidebar: Global Warming May Affect Monsoons   16
Notes   17

Chapter 2: Water, Ice, and Global Warming   19

The Ocean Absorbs Heat   19
The Ocean Absorbs Carbon   20
    Sidebar: Fight Global Warming by Putting Crop Residue in the Sea   21
Water, Climate, and Feedback Loops   22
The Impact of Water Vapor Feedback   23
The Ice Albedo Feedback Loop   24
    Graph: Arctic Sea Ice Extent   25
Other Feedback Effects   26
Notes   28

Chapter 3: Rising Sea Levels   30

Causes   30
Heated Water Takes Up More Space   30
Melting Glaciers May Raise Sea Levels   32
    Sidebar: Polar Bears and Glacial Melt   33
The Melting of Mountain Ice   34

The Melting of Giant Ice Sheets 35
    Graph: Contributors to Rising Sea Levels, 1993–2003 36
Notes 39

Debates 40
The North Atlantic Current 40
    Sidebar: Climate Disaster in Popular Culture 42
How Much Will Sea Levels Rise? 43
    Table: Rise in Sea Levels on the Atlantic Coast 47
Notes 48

Effects 48
U.S. Coastlines 49
    Table: California Population Vulnerable to a 100-Year
    Flood Along the Pacific Coast, by County 50
    Sidebar: The Wetland Feedback Loop 51
Global Coastlines 53
Islands 54
Mitigating the Damage 56
Notes 57

Chapter 4: Rainfall, Hurricanes, and Drought 59
Increased Rainfall 59
    Sidebar: Record Snowfall of 2010 Does Not
    Contradict Global Warming 61
Benefits of Increased Rainfall 62
Hurricanes and Extreme Weather Events 62
Hotter, Drier Climates 67
    Map: Projected Changes in Precipitation from 1980–
    99 to 2090–99 68
    Sidebar: Global Warming May Have Contributed to
    Slaughter in Darfur 69
Notes 72

Chapter 5: Water Supply — 74

Rivers and Water Supply — 74
Lakes and Water Supply — 75
   Sidebar: Lake Chad Disappears — 77
Melting Snow and Water Supply — 78
   Table: Predicted Change in Discharge of Selected River Mouths Due to Climate and Water Use Changes — 80
Groundwater and Water Supply — 81
Water for Crops — 82
Notes — 85

Chapter 6: Water and Energy — 87

Hydropower — 87
Tidal Power — 88
Wave Power — 90
Hydroelectric Power: Pros and Cons — 91
   Graph: United States Electrical Production in 2007 by Source of Energy — 92
Hydroelectric Power and Greenhouse Gases — 94
   Sidebar: The Vakhsh River and Climate Change — 95
Notes — 97

Chapter 7: Conclusion — 99

Water Affects Climate — 99
Sea Levels, Rain, and Drought — 101
Energy from Water — 102
The Future of Water and Climate — 103

Glossary — 104
For Further Research — 107
Index — 112
About the Author — 120

Preface

> "The warnings about global warming have been extremely clear for a long time. We are facing a global climate crisis. It is deepening. We are entering a period of consequences."
> —Al Gore

Still hotly debated by some, human-induced global warming is now accepted in the scientific community. Earth's average yearly temperature is getting steadily warmer; sea levels are rising due to melting ice caps; and the resulting impact on ocean life, wildlife, and human life is already evident. The human-induced buildup of greenhouse gases in the atmosphere poses serious and diverse threats to life on earth. As scientists work to develop accurate models to predict the future impact of global warming, researchers, policy makers, and industry leaders are coming to terms with what can be done today to halt and reverse the human contributions to global climate change.

Each volume in the Confronting Global Warming series examines the current and impending challenges the planet faces because of global warming. Several titles focus on a particular aspect of life—such as weather, farming, health, or nature and wildlife—that has been altered by climate change. Consulting the works of leading experts in the field, Confronting Global Warming authors present the current status of those aspects as they have been affected by global warming, highlight key future challenges, examine potential solutions for dealing with the results of climate change, and address the pros and cons of imminent changes and challenges. Other volumes in the series—such as those dedicated to the role of government, the role of industry, and the role of the individual—address the impact various facets of society can have on climate change. The result is a series that provides students and general-interest readers with a solid understanding of the worldwide ramifications of climate change and what can be done to help humanity adapt to changing conditions and mitigate damage.

Each volume includes:

- A descriptive **table of contents** listing subtopics, charts, graphs, maps, and sidebars included in each chapter
- Full color **charts, graphs, and maps** to illustrate key points, concepts, and theories
- Full-color **photos** that enhance textual material
- **Sidebars** that provide explanations of technical concepts or statistical information, present case studies to illustrate the international impact of global warming, or offer excerpts from primary and secondary documents
- **Pulled quotes** containing key points and statistical figures
- A **glossary** providing users with definitions of important terms
- An annotated **bibliography** of additional books, periodicals, and Web sites for further research
- A detailed **subject index** to allow users to quickly find the information they need

The Confronting Global Warming series provides students and general-interest readers with the information they need to understand the complex issue of climate change. Titles in the series offer users a well-rounded view of global warming, presented in an engaging format. Confronting Global Warming not only provides context for how society has dealt with climate change thus far but also encapsulates debates about how it will confront issues related to climate in the future.

Foreword

Earth's climate is a complex system of interacting natural components. These components include the atmosphere, the ocean, and the continental ice sheets. Living things on earth—or, the biosphere—also constitute an important component of the climate system.

Natural Factors Cause Some of Earth's Warming and Cooling

Numerous factors influence Earth's climate system, some of them natural. For example, the slow drift of continents that takes place over millions of years, a process known as plate tectonics, influences the composition of the atmosphere through its impact on volcanic activity and surface erosion. Another significant factor involves naturally occurring gases in the atmosphere, known as greenhouse gases, which have a warming influence on Earth's surface. Scientists have known about this warming effect for nearly two centuries: These gases absorb outgoing heat energy and direct it back toward the surface. In the absence of this natural greenhouse effect, Earth would be a frozen, and most likely lifeless, planet.

Another natural factor affecting Earth's climate—this one measured on timescales of several millennia—involves cyclical variations in the geometry of Earth's orbit around the sun. These variations alter the distribution of solar radiation over the surface of Earth and are responsible for the coming and going of the ice ages every 100,000 years or so. In addition, small variations in the brightness of the sun drive minor changes in Earth's surface temperature over decades and centuries. Explosive volcanic activity, such as the Mount Pinatubo eruption in the Philippines in 1991, also affects Earth's climate. These eruptions inject highly reflective particles called aerosol into the upper part of the atmosphere, known as the stratosphere, where they can reside for a year or longer. These particles reflect some of the incoming sunlight back into space and cool Earth's surface for years at a time.

Human Progress Puts Pressure on Natural Climate Patterns

Since the dawn of the industrial revolution some two centuries ago, however, humans have become the principal drivers of climate change. The burning of fossil fuels—such as oil, coal, and natural gas—has led to an increase in atmospheric levels of carbon dioxide, a powerful greenhouse gas. And farming practices have led to increased atmospheric levels of methane, another potent greenhouse gas. If humanity continues such activities at the current rate through the end of this century, the concentrations of greenhouse gases in the atmosphere will be higher than they have been for tens of millions of years. It is the unprecedented rate at which we are amplifying the greenhouse effect, warming Earth's surface, and modifying our climate that causes scientists so much concern.

The Role of Scientists in Climate Observation and Projection

Scientists study Earth's climate not just from observation but also from a theoretical perspective. Modern-day climate models successfully reproduce the key features of Earth's climate, including the variations in wind patterns around the globe, the major ocean current systems such as the Gulf Stream, and the seasonal changes in temperature and rainfall associated with Earth's annual revolution around the sun. The models also reproduce some of the more complex natural oscillations of the climate system. Just as the atmosphere displays random day-to-day variability that we term "weather," the climate system produces its own random variations, on timescales of years. One important example is the phenomenon called El Niño, a periodic warming of the eastern tropical Pacific Ocean surface that influences seasonal patterns of temperature and rainfall around the globe. The ability to use models to reproduce the climate's complicated natural oscillatory behavior gives scientists increased confidence that these models are up to the task of mimicking the climate system's response to human impacts.

To that end, scientists have subjected climate models to a number of rigorous tests of their reliability. James Hansen of the NASA Goddard Institute for Space Studies performed a famous experiment back in 1988, when he subjected a climate model (one relatively primitive by modern standards) to possible future fossil fuel emissions scenarios. For the scenario that most closely matches actual emissions since then, the model's predicted course of global temperature increase shows an uncanny correspondence to the actual increase in temperature over the intervening two decades. When Mount Pinatubo erupted in the Philippines in 1991, Hansen performed another famous experiment. Before the volcanic aerosol had an opportunity to influence the climate (it takes several months to spread globally throughout the atmosphere), he took the same climate model and subjected it to the estimated atmospheric aerosol distribution. Over the next two years, actual global average surface temperatures proceeded to cool a little less than 1°C (1.8°F), just as Hansen's model predicted they would.

Given that there is good reason to trust the models, scientists can use them to answer important questions about climate change. One such question weighs the human factors against the natural factors to determine responsibility for the dramatic changes currently taking place in our climate. When driven by natural factors alone, climate models do not reproduce the observed warming of the past century. Only when these models are also driven by human factors—primarily, the increase in greenhouse gas concentrations—do they reproduce the observed warming. Of course, the models are not used just to look at the past. To make projections of future climate change, climate scientists consider various possible scenarios or pathways of future human activity. The earth has warmed roughly 1°C since pre-industrial times. In the "business as usual" scenario, where we continue the current course of burning fossil fuel through the twenty-first century, models predict an additional warming anywhere from roughly 2°C to 5°C (3.6°F to 9°F). The models also show that even if we were to stop fossil fuel burning today, we are probably committed to as much as 0.6°C additional warming because of the inertia of the climate system. This inertia ensures warming for a century to come, simply due to our greenhouse gas emissions thus far. This committed warming introduces a profound procrastination penalty for not taking immediate action. If we are to avert an additional warming of 1°C, which would bring the net warming to 2°C—often considered an appropriate threshold for defining dangerous human impact on our climate—we have to act almost immediately.

Long-Term Warming May Bring About Extreme Changes Worldwide

In the "business as usual" emissions scenario, climate change will have an array of substantial impacts on our society and the environment by the end of this century. Patterns of rainfall and drought are projected to shift in such a way that some regions currently stressed for water resources, such as the desert southwest of the United States and the Middle East, are likely to become drier. More intense rainfall events in other regions, such as Europe and the midwestern United States, could lead to increased flooding. Heat waves like the one in Europe in summer 2003, which killed more than 30,000 people, are projected to become far more common. Atlantic hurricanes are likely to reach greater intensities, potentially doing far more damage to coastal infrastructure.

Furthermore, regions such as the Arctic are expected to warm faster than the rest of the globe. Disappearing Arctic sea ice already threatens wildlife, including polar bears and walruses. Given another 2°C warming (3.6°F), a substantial portion of the Greenland ice sheet is likely to melt. This event, combined with other factors, could lead to more than one meter (about three feet) of sea-level rise by the end of the century. Such a rise in sea level would threaten many American East Coast and Gulf Coast cities, as well as low-lying coastal regions and islands around the world. Food production in tropical regions, already insufficient to meet the needs of some populations, will probably decrease with future warming. The incidence of infectious disease is expected to increase in higher elevations and in latitudes with warming temperatures. In short, the impacts of future climate change are likely to have a devastating impact on society and our environment in the absence of intervention.

Strategies for Confronting Climate Change

Options for dealing with the threats of climate change include both adaptation to inevitable changes and mitigation, or lessening, of those changes that we can still affect. One possible adaptation would be to adjust our agricultural practices to the changing regional patterns of temperature and rainfall. Another would be to build coastal defenses against the inundation from sea-level rise. Only mitigation, however, can prevent the most threatening changes. One means of mitigation that has been given much recent attention is geoengineering. This method involves perturbing the climate system in such a way as to partly or fully offset the warming impact of rising greenhouse gas concentrations. One geoengineering approach involves periodically shooting aerosol particles, similar to ones produced by volcanic eruptions, into the stratosphere—essentially emulating the cooling impact of a major volcanic eruption on an ongoing basis. As with nearly all geoengineering proposals, there are potential perils with this scheme, including an increased tendency for continental drought and the acceleration of stratospheric ozone depletion.

The only foolproof strategy for climate change mitigation is the decrease of greenhouse gas emissions. If we are to avert a dangerous 2°C increase relative to pre-industrial times, we will probably need to bring greenhouse gas emissions to a peak within the coming years and reduce them well below current levels within the coming decades. Any strategy for such a reduction of emissions must be international and multipronged, involving greater conservation of energy resources; a shift toward alternative, carbon-free sources of energy; and a coordinated set of governmental policies that encourage responsible corporate and individual practices. Some contrarian voices argue that we cannot afford to take such steps. Actually, given the procrastination penalty of not acting on the climate change problem, what we truly cannot afford is to delay action.

Evidently, the problem of climate change crosses multiple disciplinary boundaries and involves the physical, biological, and social sciences. As an issue facing all of civilization, climate change demands political, economic, and ethical considerations. With the Confronting Global Warming series, Greenhaven Press addresses all of these considerations in an accessible format. In ten thorough volumes, the series covers the full range of climate-change impacts (water and ice; extreme weather; population, resources, and conflict; nature and wildlife; farming and food supply; health and disease) and the various essential components of any solution to the climate change problem (energy production and alternative energy; the role of government; the role of industry; and the role of the individual). It is my hope and expectation that this series will become a useful resource for anyone who is curious about not only the nature of the problem but also about what we can do to solve it.

Michael E. Mann

Michael E. Mann is a professor in the Department of Meteorology at Penn State University and director of the Penn State Earth

System Science Center. In 2002 he was selected as one of the fifty leading visionaries in science and technology by Scientific American. *He was a lead author for the "Observed Climate Variability and Change" chapter of the Intergovernmental Panel on Climate Change (IPCC) Third Scientific Assessment Report, and in 2007 he shared the Nobel Peace Prize with other IPCC authors. He is the author of more than 120 peer-reviewed publications, and he recently coauthored the book* Dire Predictions: Understanding Global Warming *with colleague Lee Kump. Mann is also a cofounder and avid contributor to the award-winning science Web site RealClimate.org.*

CHAPTER 1

Water and Climate: An Introduction

Water covers seventy percent of the earth's surface. As a result, the earth's climate is profoundly affected by water in all of its forms. Indeed, much of what is perceived as climate—the amount of rain year round, the amount of snow year round, the level of humidity—is simply another way of talking about the amount and kind of water in the air. Other aspects of climate, such as temperature, are also profoundly affected by water.

The Hydrologic Cycle

One of the main ways water affects climate is through the hydrologic cycle. The hydrologic cycle is "the pilgrimage of water as water molecules make their way from the earth's surface to the atmosphere and back again."[1] In the hydrologic cycle, water is constantly moving between different states (solid, liquid, and gas) and different sites on the earth (the ocean, the atmosphere, and the land).

As an example of how the hydrologic cycle works, imagine a glass of water outside a home on a warm day. Some of the water in that glass will evaporate and go into the atmosphere as water vapor. It may stay in the atmosphere for a while, and then eventually fall as rain into a nearby lake. That lake may in turn supply that same home with water. Eventually, when the faucet is turned on, the very same water molecules that were in that glass of water may be reused.

Water and Climate: An Introduction

The transformation of water into vapor and back to water—the hydrologic cycle—is powered by the sun. It is the sun that provides the heat energy that turns liquid water into vapor. Specifically, "Heating of the ocean water by the sun is the key process that keeps the hydrologic cycle in motion."[2] Most of the water in the atmosphere—as much as 90 percent—comes from evaporation from the oceans and other large bodies of water, such as lakes.[3] A smaller amount comes from sublimation, which occurs when, near the freezing point at certain pressures, snow or ice turns directly into water vapor without first turning into water.

> *According to the NASA Web site* Earth Observatory, *"A cornfield 1 acre in size can transpire as much as 4,000 gallons of water every day." That is about as much water as would be needed for a 19-hour shower.*

A significant amount of water vapor in the air comes from plants. Plants take in water through their roots and let it out through pores called stomates on their leaves. This release of water from plants is called transpiration. According to the NASA Web site *Earth Observatory*, "a cornfield 1 acre in size can transpire as much as 4,000 gallons of water every day."[4] That is about as much water as would be needed for a 19-hour shower.

Once water is in the atmosphere, it rises, cools, and condenses, turning into clouds and eventually falling back to the earth. The water may fall directly into the ocean, or it may fall on land, to be intercepted by plants, or to percolate into the ground, or to end up in streams or lakes or other bodies of water. Eventually, the water evaporates again, and the cycle continues.

The water cycle is broadly the same everywhere: Water turns to vapor and rises into the air; vapor turns to water and falls to the earth. Individual variations in the cycle have an enormous effect on local climates, however. For example, around the Great Lakes in winter, large amounts of water evaporate from the lakes,

Water and Ice

LOCATION AND FORM OF WATER ON EARTH

Water Source	Percent of Total Water
Oceans, Seas, and Bays	96.5
Ice Caps, Glaciers, and Permanent Snow	1.74
Groundwater	1.7
Fresh	0.76
Saline	0.94
Soil Moisture	0.001
Ground Ice and Permafrost	0.022
Lakes	0.013
Fresh	0.007
Saline	0.006
Atmosphere	0.001
Swamp Water	0.0008
Rivers	0.0002
Biological Water	0.0001
TOTAL	**100**

Source: "Where Is Earth's Water Located?" *USGS Water Science for Schools*, July 15, 2009. http://ga.water.usgs.gov.

and this moisture is blown toward shore as clouds. "By the time these clouds reach the shoreline, they are filled with snowflakes too large to remain suspended in the air and, consequently, they fall along the shoreline as precipitation," according to the Web site *WW2010* maintained by the University of Illinois at Chicago.

The precipitation that falls near the lake is often referred to as lake-effect snow, and it can be extremely heavy. For instance, in November 1996, Cleveland, Ohio, received more than 50 inches (127cm) of snow over the course of two to three days due to the lake effect.[5]

As the lake effect demonstrates, variations in the hydrologic cycle can have substantial effects on the amount of precipitation. Later chapters will look at how global warming may affect the water cycle and rainfall.

The Ocean Affects Climate

Even outside its role in the hydrologic cycle, the ocean has a major effect on climate. In particular, oceans can have a major impact on temperature, in part because water has a high specific heat. Specific heat is a measure of the amount of energy needed to increase the temperature of a substance by a set amount. So to say that water has a high specific heat is simply to say that it heats up (and cools down) more slowly than many other substances.

The high specific heat of water should be familiar from many everyday situations. For example, on the beach on a very hot day, the sand can burn a person's feet even when the water is relatively cool. This is true in part because water heats up more slowly than sand. Furthermore, in winter, lakes often remain liquid even when the ground is frozen because the water cools down more slowly than the land.

Water, then, essentially resists changes in temperature—compared to the land, water is warmer in winter and colder in summer. This has a major effect on the climates near the ocean, called maritime climates. Maritime climates are usually mild; they do not get too cold in winter, nor do they get too warm in summer. Maritime climates are also often humid, as the winds off the ocean carry water vapor that cools and falls to the earth when it blows over land. The British Isles are an example of a maritime climate, with temperatures that vary only about 50°F (28°C) over the course of a year.[6]

Continental climates, on the other hand, are climates usually found within the interior of large landmasses. Continental climates have hot summers and cold winters. They are often relatively dry, because water from the ocean often falls to the earth before it reaches continental interiors. Siberia in Russia is an example of a continental climate. Temperatures in Siberia can vary as much as 104°F (58°C) between winter and summer.[7]

The way winds interact with the ocean also have an important impact on climate. For example, along the eastern seaboard of the United States, winds tend to move from west to east. Thus, because the wind is coming across the continent, rather than off the ocean, states such as New York and Massachusetts do not have a maritime climate even though they are close to the ocean. Instead, they have what is essentially a continental climate.

Water, Climate, and People

The interaction of water and climate has a major impact on human life and health. The hydrologic cycle, for example, provides precipitation that fills reservoirs and lakes, thus providing drinking water. Climate patterns influenced by the ocean determine where people live and have major effects on economic activity, quality of life, and every other aspect of human behavior.

The interaction between water, climate, and people is complicated. On one hand, changes in water and climate can affect people. On the other hand, people can in turn affect water and climate. One example of this interrelationship can be seen in connections between human development and the hydrologic cycle. When human beings build cities, they pave large areas of land. Water that used to soak into soil and grass cannot soak through pavement. As a result the growth of cities, or urbanization, "generates more stormwater runoff and higher peak stream discharge."[8] Peak stream discharge is the amount of water passing through any part of a stream at a given time. When storm water runoff and peak stream discharge are high, it means that more water is running faster into and down a stream, which can

Water and Climate: An Introduction

Changes to the El Niño weather system have caused more hurricanes and flooding, such as this flood in Bolivia in 2007. Although El Niño is probably natural, the recent changes may be a result of global warming. Aizar Raldes/AFP/Getty Images.

result in more flooding. Thus, human development can have a strong effect on the hydrologic cycle, and those changes in the water cycle can in turn have a large, and sometimes negative, effect on human quality of life. Building cities in some areas can cause floods that damage those cities and the people who live in them.

Another larger and more complicated example of the interconnection of people, water, and climate is El Niño. El Niño is a periodic warming of the ocean. This warming appears to be natural, and it can have major effects on weather. Traditionally, El Niño events have occurred in the Pacific, close to the South American coast. Such El Niño events usually mean a quieter storm season. Recently, however, El Niño events have been occurring farther west in the central Pacific. These new El Niños have been "resulting in a greater number of hurricanes with greater frequency and more potential to make landfall," accord-

Water and Ice

Global Warming May Affect Monsoons

One of the most important and striking climate phenomena in the world is the monsoon. A monsoon is a seasonal prevailing wind that lasts for several months. In India and surrounding countries, a monsoon blowing from the Indian Ocean results in torrential rains between June and September. In India, around 35 inches (89cm) of rain usually fall in that period.

A study led by Noah Diffenbaugh at Purdue University's Climate Change Research Center suggests that global warming may push back the start of the monsoon season in India by 5 to 15 days. At first, a delay in the monsoons may seem like a good thing: The downpour causes flooding and even landslides. India relies on the rainfall, however, to provide water for drinking, for agriculture, and for hydroelectric power. If the monsoons are delayed, it could have serious negative consequences for huge numbers of people.

A reduction in monsoon winds where they are expected can be bad. But the appearance of monsoon winds where they are not anticipated could also cause problems. University of Oregon Geology Professor Greg Retallack has studied historical changes in climate patterns, and he believes that if world temperatures warm, the Pacific Northwest may begin to experience monsoon storms. In one global warming event 55 million years ago, rainfall increased by 50 percent.

If warming again triggered monsoons on that scale, the climate and perhaps the landscape of Seattle and the surrounding region would be significantly changed, resulting in major adjustments for people living there.

Changes in monsoon patterns demonstrate again that water and climate can be connected in complicated ways. Thus, global warming may change weather patterns around the globe in ways that are hard to predict but may have major consequences for millions of people.

ing to Peter Webster of Georgia Tech's School of Earth and Atmospheric Sciences.[9]

The changes in El Niño may be in part a result of global warming. Or they might not be. Webster, who was co-author of a study investigating the El Niño effect, was not sure. However, the possibility that global warming may change water temperatures, which could in turn affect the frequency and power of storms, has led to both concern and debate among scientists. University of Chicago ocean chemist David Archer notes that "Tropical storms tend to intensify as the sea surface warms, everything else being equal. Hurricane intensities reconstructed from satellite images seem to show strengthening as sea surface temperature warms over the past decades." Archer also notes that "though there is a clear danger that hurricanes could intensify. . . . Scientists don't understand the working of hurricanes in the climate system well enough to be able to predict how strong they will get."[10]

Thus, scientists know there is a link between water temperature and storm systems, but how those links work exactly, and how they may affect people, are difficult to determine. Scientists do know, however, that as the earth's temperature increases, the changes may affect the earth's water in many different, surprising, and possibly dangerous ways. In turn, the changes in the oceans, in precipitation, in the icecaps, and in lakes and rivers may have powerful effects on climate and on human beings.

Notes

1. Steve Graham, Claire Parkinson, and Mous Chahine, "The Water Cycle," *Earth Observatory*, n.d. http://earthobservatory.nasa.gov.
2. *Environment Canada*, "The Hydrologic Cycle," July 25, 2008. www.ec.gc.ca.
3. Graham et al., "The Water Cycle."
4. Graham et al., "The Water Cycle."
5. WW2010, "A Summary of the Hydrologic Cycle," n.d. http://ww2010.atmos.uiuc.edu.
6. "Maritime Climate," *Atmosphere, Climate & Environment Information Programme*, n.d. www.ace.mmu.ac.uk.
7. "Continental Climate," *Atmosphere, Climate & Environment Information Programme*, n.d. www.ace.mmu.ac.uk.
8. *Chesapeake Bay & Mid-Atlantic from Space*, "What Are the Effects?" n.d. http://chesapeake.towson.edu.

9. Doyle Rice, "El Niño 2 Fuels More Atlantic Hurricanes, Warnings," *USA Today*, July 13, 2009. www.usatoday.com.
10. David Archer, *The Long Thaw: How Humans Are Changing the Next 100,000 Years of Earth's Climate*. Princeton, NJ: Princeton University Press, 2009, pp. 48–49.

Chapter 2

Water, Ice, and Global Warming

As noted in the last chapter, water has a relatively high specific heat, which means that it heats up and cools down more slowly than many other substances. Without water, the earth would be much hotter. As Robert Stewart, a professor of oceanography and meteorology at Texas A&M says, "For insights into what Earth might be like if there had never been an ocean, hop over to Venus, where a runaway greenhouse effect has rendered the surface hot enough to melt lead."[1] The ocean, in other words, moderates temperatures. Thus, for example, in Chicago, temperatures close to the lake are generally slightly cooler in summer and slightly warmer in winter than temperatures farther from the water. Similarly, as discussed in the last chapter, people who live along a coast are more likely to experience moderate temperatures than those who live at the same latitude in the middle of a continent.

The Ocean Absorbs Heat

Just as water moderates seasonal changes in temperature, so does the ocean act as a brake on climate change. The earth's water, in liquid form in the oceans and in the form of ice sheets at the poles, acts as a giant heat sink. Energy pours into the oceans and is held there, where it cannot heat up the land. According to University of Chicago ocean chemist David Archer, "the oceans and ice sheets hold so much heat that they take thousands of years to respond fully to changing climate."[2]

Water and Ice

The earth's water does much more than simply slow down climate change, however. The ocean redistributes heat throughout the world in myriad ways. For example, water at the surface of the ocean constantly evaporates into water vapor, which rises into the atmosphere. Eventually, the heat energy is released through condensation, when the water vapor turns back into water and falls as precipitation. This heat cycle is what powers the world's winds, and thus much of the earth's climate. As Robert Stewart notes, "In the tropics in particular, where energy from the sun is at its greatest, the exchange of heat between ocean and atmosphere drives much of the global atmospheric circulation."[3] Ocean currents also redistribute heat by, for example, pushing warm water from the equator north, and thereby moderating temperatures at high latitudes. The effects of global warming on precipitation will be discussed more fully in later chapters.

The Ocean Absorbs Carbon

In addition to absorbing and distributing heat, the ocean also affects global warming by absorbing carbon dioxide. Carbon dioxide (CO_2) is the main way in which humans contribute to global warming. The burning of fossil fuels puts a great deal of carbon dioxide into the atmosphere, trapping the sun's heat and raising temperatures on the earth. By taking carbon dioxide out of the atmosphere, therefore, the oceans slow the process of global warming. According to David Adam, the environmental correspondent for *The Guardian*, "The world's oceans soak up about 11bn [billion] tonnes of human carbon dioxide pollution each year, about a quarter of all produced."[4]

There is some evidence, however, that as the earth warms, the ocean may become less effective at absorbing CO_2 from the atmosphere. There are numerous reasons for this. In the first place, warming of the oceans seems to disrupt the circulation and mixing of surface and deep water. Carbon in deep water is more likely to remain captured, and if carbon is not dragged down to deep water, water at the surface will become saturated

Fight Global Warming by Putting Crop Residue in the Sea

Plants such as corn contain a great deal of carbon. Keeping that carbon from entering the atmosphere as carbon dioxide could greatly reduce greenhouse gas emissions. Two scientists, Stuart Strand and Gregory Benford have a simple suggestion for how to do that. Simply take crop residue—corn husks, plant stalks, and other plant waste—bundle it up, and drop it into the ocean.

Traditionally, environmentalists have tried to find other green uses for crop residue—using it for fuel, for example. But Strand and Benford argue that dumping it in the ocean is the easiest and best solution. Even with the cost of bundling and transporting, Strand and Benford found, dropping the crops in the ocean reduces global carbon dioxide emissions by 15 percent.

There are a couple of problems with the proposal. The most serious is probably that Strand and Benford assume that the crop residue will be dropped into deep water, where it would keep the carbon from turning into carbon dioxide for millennia. However, other studies suggest bacteria and other organisms may break down the plant matter long before then, with carbon dioxide as a byproduct.

In addition, many researchers still hope that as fossil fuels run out or become more expensive, biofuels—fuels made from plant matter—will eventually become popular. If this happened, it might then be much more productive to use a corn stalk to power a car than it would be to sink that same stalk to the bottom of the sea.

Nonetheless, as carbon dioxide levels rise, and if other plans to reduce emissions fail to gain traction, some nations may try dropping stalks in water, however silly it may sound.

more quickly and unable to absorb more carbon dioxide. Thus, when ocean circulation is weakened, less carbon is deposited in deeper waters, and the amount of carbon absorbed by the ocean overall is reduced. A study led by Kitack Lee of South Korea's

Pohang University of Science and Technology found that carbon uptake (that is, the amount of carbon absorbed) in the Sea of Japan from 1999 to 2007 was half that of carbon uptake from 1992 to 1999. The scientists attributed this effect to circulation changes caused by global warming.[5]

The chemistry of the ocean may also change as CO_2 in the atmosphere increases, and this reaction may, in turn, limit carbon dioxide absorption. The ocean currently is filled with charged particles called carbonate ions. These carbonate ions react with CO_2 in the atmosphere and turn it into the molecule bicarbonate, which cannot evaporate again into the air. David Archer argues that as more CO_2 is pumped into the atmosphere, the ocean will start to run out of carbonate ions. The result will be that the more carbon there is in the atmosphere, the more slowly the ocean will absorb carbon. "The bottom line," Archer says, " . . . is that the natural world takes up fossil-fuel CO_2 more slowly than we might have expected given how much CO_2 is dissolving in the oceans today."[6]

Water, Climate, and Feedback Loops

The cycle, whereby more carbon dioxide in the atmosphere reduces the absorptive power of the ocean, thus increasing the amount of carbon dioxide in the atmosphere, is an example of a feedback loop. In general, a feedback loop is a situation in which one condition creates other conditions that reinforce the first.

For example, imagine a student is late and hurrying to get out of the house, so he accidentally drops his keys in the toilet. Panicked, he stumbles and hits the handle, flushing the keys away. Even more panicked, he races around looking for the spare keys, trips, and twists his ankle. More panicked than ever, he stumbles about . . . and so forth. The fact of being late creates other conditions (loss of keys) that make the student fall even further behind.

Scientists believe that numerous feedback loops exist that may result from the interaction of global warming and water.

One of the most powerful involves water vapor. Water vapor is the gaseous form of water. It is also what physicist John Cook calls "the most dominant greenhouse gas."[7] Water vapor is a significant component of the atmosphere, and it is much more effective at trapping the sun's rays than is carbon dioxide. Whereas carbon dioxide levels can rise more or less indefinitely, however, there is a limit to how much water vapor the atmosphere can hold before releasing it through precipitation.

Nonetheless, water vapor can have a significant effect on climate change. As temperatures go up, more water evaporates from the surface of the ocean. That means more water vapor is in the atmosphere, which traps more of the sun's rays, and thus increases global warming. This is, then, a classic feedback loop: Warming produces water vapor, which adds to the warming.

The Impact of Water Vapor Feedback

The Web site *Skeptical Science* argues that the water vapor feedback effect multiplies by two the impact of global warming. In other words, without water vapor feedback, doubling the amount of CO_2 in the atmosphere would raise global temperature by about 1.8°F (about 1°C). Counting the water vapor feedback effect, the temperature would rise by about 3.6°F (about 2°C).

Some debate exists about whether water vapor has such a strong feedback effect, however, and about how or whether this effect might increase or decrease as the earth warms further. David Archer notes that the earth currently has an average relative humidity of about 80 percent—relative humidity being the amount of water vapor that exists in the atmosphere. Archer suggests that an increase in global temperature and global CO_2 might increase the global average relative humidity, possibly resulting in a larger feedback effect. On the other hand, if increases in temperature cause the relative humidity to drop, the feedback effect might be smaller than expected.

One of the most controversial arguments about water vapor feedback has to do with clouds. When water evaporates, it causes clouds to form. Clouds have a high albedo—which is to say, they reflect sunlight rather than absorbing it. A lot of clouds would therefore increase the albedo of the earth. With a higher albedo, the earth would reflect more sunlight, and global temperatures would cool. Science educator Randy Russell has noted therefore that "an increase in water vapor, and hence cloudiness, might actually serve as a 'self-correcting' mechanism (or 'negative feedback loop') that would 'put the brakes on' global warming; or possibly induce a period of 'global cooling.'"[8]

> *Summer sea ice in the Arctic has fallen by about 386,102 square miles (about one million square kilometers) in the past 30 years.*

It is not at all clear how the water vapor feedback loop and the cloud albedo feedback loop interact. As just one complicating factor, consider the idea that global warming may change regional climates and precipitation patterns in unexpected ways, which in turn could have a powerful effect on humidity and cloud formation. Because of the difficulty of the problem, and because of the importance of both of these loops to overall climate, Russell argues that "Predicting the net influences these feedback loops produce is possibly the greatest challenge facing modern climate scientists who are trying to determine our future climate."[9]

The Ice Albedo Feedback Loop

Another important feedback loop is called the ice albedo effect. As noted above, albedo is the ability of a substance or a body to reflect sunlight. Ice and snow are white, which means that they reflect the sun's rays—or, to put it another way, they have very high albedos. In fact, according to geographer Grant R. Bigg, ice and snow have "the highest albedos of almost any natural surface."[10] From 70 percent to 80 percent of the sunlight that strikes ice and snow is reflected back into space.

Water, Ice, and Global Warming

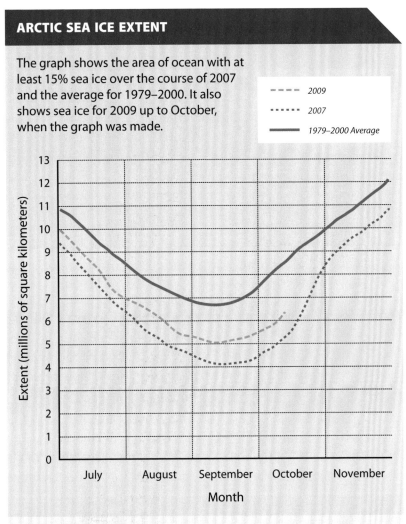

ARCTIC SEA ICE EXTENT

The graph shows the area of ocean with at least 15% sea ice over the course of 2007 and the average for 1979–2000. It also shows sea ice for 2009 up to October, when the graph was made.

Source: "Arctic Sea Ice Extent," National Snow and Ice Data Center, October 12, 2009. http://inside.org.

The ice albedo effect is most important at the poles where large expanses of ice and snow reflect sunlight and help to keep the climate cold. Because of the ice albedo effect, relatively small shifts in temperature at the poles may cause major climatic fluctuations. For instance, some scientists believe that in the past

Water and Ice

when temperatures dropped, it created more ice cover, which reflected more sunlight, lowering temperatures even further. There was a period 800 million years ago, in particular, when some scientists believe that runaway ice albedo feedback was responsible for a remarkable period of cooling, during which ice formed all over the globe, even near the equator.

With global warming, the earth may be set to experience a feedback loop of the opposite kind. As the earth's temperature starts to rise, ice and snow at the poles will start to melt. Land and water have a much lower albedo than ice: for example, ocean water reflects less than 10 percent of the sunlight that strikes it. As more sunlight is absorbed, temperatures increase—which, in turn, melts more ice and further increases the temperature.

Scientists worry that ice albedo feedback is in part responsible for the fact that summer sea ice in the Arctic has fallen by about 386,102 square miles (about one million square kilometers) in the past 30 years. Environmental education specialist Lisa Gardiner has noted that "According to climate models, the pace of ice melt will continue to quicken so much that that there may be no more summer sea ice within the next few decades."[11]

Other Feedback Effects

Scientists wonder about several other types of feedback effects as well. For example, as temperatures warm, frozen ground in the Arctic will thaw. Partially decayed vegetation, or peat, which has been locked in the permafrost, will start to decay and release carbon dioxide into the atmosphere. This, in turn, might exacerbate global warming. At the same time, less permafrost would mean less frozen ground, and more places for trees to grow. Trees take in carbon dioxide, so more trees might have the opposite effect, reducing the amount of CO_2 in the atmosphere and slowing global warming.

Or, as another example, it is expected that global warming would cause hotter, drier weather in some areas. This change might cause an increase in forest fires. Burning wood releases

Water, Ice, and Global Warming

Global warming may create a feedback loop in which reduced polar ice reflects less sunlight, which leads to greater warming and more polar ice melting. DEA/C. Sappa/De Agostini/Getty Images.

carbon dioxide into the air, which might enhance the greenhouse effect. On the other hand, greater dryness might lead to desertification in some places. As land turned into desert, more windblown dust would enter the earth's atmosphere. This dust could block sunlight, lowering temperatures and reducing the greenhouse effect.

On one hand, feedback effects demonstrate how interconnected climate and water are, and how difficult it can be to know what even small changes in temperature may do, or may not do, to climate worldwide. The danger of feedback effects has led many scientists to point to the necessity for quick action. For instance, James Hansen, director of NASA's Goddard Institute for Space Studies in New York, and a leading proponent of action on climate change has said that "strong amplifying feedbacks" may cause the earth to pass "dangerous tipping points."[12]

On the other hand, some kinds of feedback loops may be overestimated. For example, some scientists have suggested that an increase in global temperatures may cause soils to release more carbon dioxide into the atmosphere, resulting in a feedback loop and even more warming. Cornell professor of biogeochemistry Johannes Lehmann studied the amount of black carbon in Australian soils. What he discovered was that black carbon actually stayed in soil much longer than researchers had originally believed. Scientists, in other words, had overestimated the feedback effect—a rise in temperature would not cause carbon to escape quickly from Australian soils, and so there would be no feedback effect. Thus, Lehmann noted that though climate change is a serious problem, "this particular aspect, black carbon's stability in soil, if incorporated in climate models, would actually decrease climate predictions" of warming.[13]

Because feedback effects can exert an unusually large influence on climate, understanding them is vital to predicting and dealing with climate change. Scientists believe they have made progress in identifying some of the most important areas of concern, such as the ice albedo effect and the water vapor feedback loop. Still, researchers are a long way from understanding all the ways in which water and climate may interact.

Notes

1. Robert Stewart, "Ocean and Climate," *Ocean Motion*, n.d. www.oceanmotion.org.
2. David Archer, *The Long Thaw: How Humans Are Changing the Next 100,000 Years of Earth's Climate*. Princeton, NJ: Princeton University Press, 2009, p. 68.
3. Stewart, "Ocean and Climate."
4. David Adams, "Sea Absorbing Less CO_2, Scientists Discover," *Guardian*, January 12, 2009. www.guardian.co.uk.
5. Adams, "Sea Absorbing Less CO_2, Scientists Discover."
6. Archer, *The Long Thaw*, p. 113.
7. *Skeptical Science*, "Water Vapor Is the Most Powerful Greenhouse Gas," n.d. www.skepticalscience.com.
8. Randy Russell, "Global Warming, Clouds, and Albedo: Feedback Loops," *Windows to the Universe*, May 17, 2007. www.windows.ucar.edu.
9. Russell, "Global Warming, Clouds, and Albedo: Feedback Loops."
10. Grant Bigg, *The Ocean and Climate*, 2nd ed. Cambridge, UK: Cambridge University Press, 2003, p. 238.
11. Lisa Gardiner, "Ice-Albedo Feedback: How Melting Ice Causes More Ice to Melt," *Windows to the Universe*, July 18, 2007. www.windows.ucar.edu.

12. Quoted in Bill Blakemore, "NASA: Danger Point Closer than Thought from Warming," *ABC News Online*, May 29, 2007, p. 1. www.abcnews.go.com.
13. Quoted in Krishna Ramanujan, "Soil Study Suggests Future Climate Change Models Should Be Revised," *Cornell Chronicle*, November 18, 2008. www.news.cornell.edu.

Chapter 3

Rising Sea Levels

CAUSES

Scientists surmise that global warming may cause a significant rise in sea levels. At first this may seem backwards. After all, as temperatures go up, one might think there would be more evaporation, and that the oceans would recede. So why does the opposite happen?

Heated Water Takes Up More Space

Scientists believe ocean levels will rise for several reasons. The first is thermal expansion. When objects are heated, they tend to expand. People regularly use this principle in the kitchen. If there is a lid stuck on a jar, one way to remove it is to run the lid under hot water. When the lid is heated, it expands so that it can be easily removed (the glass jar expands too, but at a slower rate).

When water is heated, it expands and takes up more space. Thus, if global temperatures rise, the ocean will heat up, and the water in the ocean will take up more space. The Intergovernmental Panel on Climate Change (IPCC) "projects that thermal expansion will be the main component of expected sea-level rises over the 21st century."[1]

Not all scientists agree that global warming will result in the thermal expansion of water. Marcel Leroux, a French climatologist and a prominent opponent of global warming theories before his death, claimed that thermal expansion of seawater was not well documented. Leroux argued that if seawater expands with changes in temperature, such expansion should be apparent with the changes in the seasons. In other words, sea levels should be higher in the summer, when it is hot, than in the winter, when it is colder. Working with data from the tide-gauge at Brest in France, Leroux concluded that "variations in water levels and temperatures are not synchronous at Brest; the highest (positive) values do not occur in summer, but in autumn-winter."[2] In other words, sea levels were higher in winter than in summer, which is the opposite of what one would expect if thermal expansion were occurring. Leroux does not explain why sea levels were higher in winter, and there might be any number of reasons for it, but the fact remains that his data did raise some questions about thermal expansion.

Another researcher who discovered data that seemed to contradict thermal expansion was Josh Willis, a scientist at NASA's Jet Propulsion Laboratory. Willis works on estimating the amount of heat in the ocean. In 2004, Willis published a study showing that the ocean heat rose from 1993 to 2004. In 2006, he did a follow-up study—but much to his surprise, the research now showed the ocean cooling from 2003 to 2005. This result was especially confusing because other data showed that ocean levels during that period were rising. Willis himself believed that the data simply demonstrated natural variation in temperature. Others though "cited the results as proof that global warming wasn't real and that climate scientists didn't know what they were doing."[3]

In 2007, however, Willis discovered that he had been in error. The ocean cooling he had discovered did not exist. His data had been bad, and in fact the ocean had warmed between 2003 and 2005. The revised data turned out to be very important in calculating sea-level rise. According to Catia Domingues, a scientist with Australia's Commonwealth Scientific and Industrial Research Organisation, using Willis's new data allowed researchers to see "that ocean heating was larger than scientists previously thought, and so the contribution of thermal expansion to sea level rise was actually 50 percent larger than previous estimates."[4]

Melting Glaciers May Raise Sea Levels

Another important cause of sea-level rise is ice melt. The term "ice melt" may well evoke an image of floating icebergs dissolving into the sea and raising the level of the ocean. In fact, however, icebergs are the one kind of ice that scientists do *not* worry about when they think of rising sea levels. That is because icebergs are already floating. Just as the water level in a glass rises when ice cubes are dropped into it, so the ocean level is *already* higher because it has icebergs floating in it.

Icebergs melting can have very important effects. In particular, sea ice has a much higher albedo than water. As discussed in the previous chapter, if too much sea ice melts, there will not be as much ice to reflect the sun's rays, and the overall temperature of the ocean will rise. But these effects do not have any direct effect on ocean level. In fact, if you melted all the sea ice on the earth, the ocean level would not rise at all, just as, if you melted the ice in a glass of water, the level of water in the glass would not change.

When scientists think about sea-level rise, therefore, they are contemplating glaciers, rather than icebergs. Glaciers are basically large masses of ice on land. They are created in areas where much more snow falls in the winter than can melt in the summer. Compressed under its own weight over many years, the snow eventually turns into ice.

Glaciers can be enormous: some are more than a hundred miles long. "Almost 10 percent of the world's landmass is currently covered with glaciers, mostly in places like Greenland and Antarctica."[5] There are also glaciers at high altitudes on the tops of mountains. Altogether, glaciers contain about 75 percent of the world's freshwater supply.

If a glacier melted and turned into water, and that water flowed into the ocean, then the ocean level would in fact rise. But

Polar Bears and Glacial Melt

The melting of ice as global temperatures rise has effects not only on humans, but on other species as well. One animal that has been especially hard hit by the decrease of ice is the polar bear.

Polar bears typically hunt in water, often surfacing to cling to ice floes and then swim to land. Although bears can swim 100 miles without much trouble, "as the ice gets farther out from shore because of warming, it's a longer swim that costs more energy and makes them more vulnerable," according to Ian Stirling of the Canadian Wildlife Service as quoted by Bill Mouland in the *Daily Mail* on February 1, 2007.

In addition, polar bears' winter shelters in snowdrifts and on sea ice have been disappearing as the temperature warms. As their habitat diminishes, polar bear numbers have declined, dropping by a quarter in the past twenty years. In addition, individual polar bears have become smaller and thinner. Females especially have diminished in size. Partially as a result, polar bear litters have shrunk; triplets, which used to be fairly common, are now unheard of.

With current predictions of global temperature increases and continued northern ice melt, the outlook for polar bears seems grim. Some scientists hope that special conservation areas may help the bears survive. If not, extinction in the near future seems like a real possibility.

measuring the exact amount of glacial melt, and its contribution to sea-level rise, is not easy to do. As just one problem, "Neither the USA nor Canada has completed a national glacier inventory," making it impossible to collect a complete set of data, according to Roger Braithwaite, who worked on a study of glacial melt for the Alfred Wegener Institute for Polar and Marine Research.[6] Despite this and other difficulties, however, scientists have estimated that glacial melting seems to be raising the sea level by about .4 inches (1mm) a year, according to University of Victoria climate scientist Andrew Weaver. Weaver noted that, "These numbers may not seem like much in any given year, but if you consider changes over longer timescales, they start to add up."[7]

The Melting of Mountain Ice

Scientists agree that global warming will cause glacial melt, and that glacial melt will contribute to sea-level rise. Nonetheless, there remain numerous disagreements about the exact process and extent of glacial melting and its relation to global warming. As just one example, the most famous melting ice in the world may be the glacier atop Mount Kilimanjaro in Tanzania. The icecap on Mount Kilimanjaro has declined visibly over the last hundred years, and this phenomenon is often blamed on global warming. For example, Lonnie Thompson, an Ohio State University researcher of ancient climates, has linked the Kilimanjaro melt to global warming. Thompson said that Kilimanjaro had lost 25 percent of its ice between 2000 and 2006. In speaking about a number of glaciers, including Kilimanjaro, he said that "they're not just retreating, they're accelerating [their retreat]. . . . And it raises the question of whether this might be a fingerprint of [human-caused global] warming."[8] Further research as of 2009 led Thompson to link the melting ice even more clearly to global warming, and to predict that the glaciers atop Kilimanjaro could disappear entirely by 2034.[9]

Together, the Antarctic and Greenland ice sheets contain enough water to raise the level of the ocean by more than 223 feet (68m).

Other researchers disagree that the melt of Kilimanjaro is caused by global warming, however. For example, researchers Philip Mote of the University of Washington and Georg Kaser of the University of Innsbruck argue that Kilimanjaro's glacial melt was well under way in the first half of the 1900s, before global warming would have been an important factor. Instead,

Mote and Kaser blame other factors, such as a decline in regional snowfall. Mote noted in 2007 that there were many other glaciers about which there was "absolutely no question that they are declining in response to the warming atmosphere." The glacier on Kilimanjaro simply happened not to be among them.[10]

It is important to note that mountain glacial melt can have a number of important effects beyond raising sea levels. The melting of the glacier on Kilimanjaro, caused by global warming or not, may damage the mountain's attractiveness as a tourist destination, with dramatic economic consequences for Tanzania. More broadly, the melting of mountain glaciers can result in the creation of huge lakes and dangerous floods. Flooding in Yosemite National Park in 1997 and 2005, and a devastating flood in Bhutan in 1994 were all linked to glacial melting. Henry Diaz of the U.S. Office of Oceanic and Atmospheric Research warned that with global warming "These kinds of things are likely to occur more frequently in the future."[11]

The Melting of Giant Ice Sheets

In thinking about the effect of ice melt on sea levels, scientists are especially concerned about two massive glaciers known as ice sheets, one in Greenland and one in Antarctica. The Greenland ice sheet covers .65 million square miles (1.7 million sq. km), and contains .68 million cubic miles (2.85 million cubic km) of ice. The Antarctic ice sheet is many times larger; it covers more than 4.6 million square miles (12 million sq. km), and contains more than 6 million cubic miles (25 million cubic km) of ice. Together, these ice sheets contain enough water to raise the level of the ocean by more than 223 feet (68m) according to a Working Group contribution to the 2001 Intergovernmental Panel on Climate Change report.[12] The ice sheets, in other words, contain massive amounts of water. If global warming caused them to melt to any significant degree, sea levels could rise significantly.

For the short term, according to ocean chemist David Archer, "The current state-of-the-art computer models of ice sheets pre-

Water and Ice

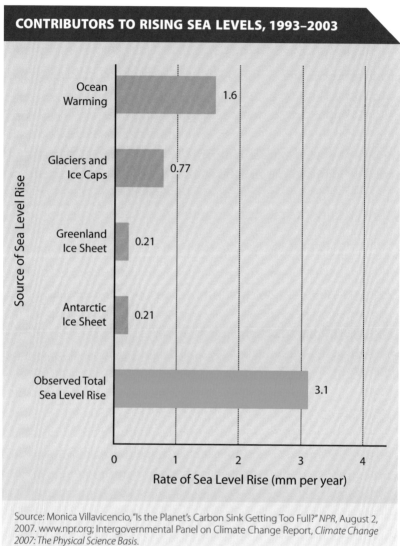

dict that the ice sheets will not melt into the ocean very much in the next hundred years." Archer also notes, however, that "there are examples from the past of ice sheets collapsing into the ocean over only a few centuries . . . ice sheets may know a few tricks about melting that the ice sheet models have still to figure out."[13]

As one example of such a "trick," Archer points to the sudden break up, or as it is called, the explosion of the Larsen B ice shelf in the Antarctic Peninsula in 2002. An ice shelf is a thick, floating platform of ice that forms at the edge of a glacier. The 2002 explosion changed "a continuous region of ice the size of New Hampshire into a blue slurpy mash of tiny icebergs in just a few days."[14] Scientists believe this sudden explosion may be related to an increase in ponds of meltwater forming at the surface of the ice. Such ponds have also been observed on the surface of other ice shelves, and scientists are concerned that similar explosions may occur in the future.

Ice shelves are already floating on water, so their breakup does not directly affect sea levels. The disappearance of the ice shelf does seem to increase the rate of seawater melt from adjacent land-based ice, however. Dramatic events such as ice-shelf explosions also emphasize the way in which feedback events can cause significant and unpredictable melting.

Scientists worry about several possible feedback effects that may cause the ice sheets to melt more quickly than anticipated. The first is the albedo effect, mentioned in the previous chapter and above in the discussion of sea ice melt. Again, if sea ice melts, it does not raise sea levels. But it will lower albedo, as water reflects far less light than ice does. This effect is of particular concern in the waters around Greenland, where there are significant amounts of floating ice. As that sea ice melts, the region will warm, and the Greenland ice sheet may also melt.

Another feedback mechanism that worries some scientists also involves melting ice. "As an ice sheet's surface begins to melt, some of the water filters down through cracks in the glacier, lubricating the surface between the glacier and the rock beneath it. This accelerates the glacial flow and the calving [breaking away] of icebergs into the surrounding ocean," according to environmentalist and founder of the Earth Policy Institute Lester R. Brown.[15] Brown also argues that such water, as it flows into the glacier, carries heat to lower parts of the glacier faster than might be expected.

Water and Ice

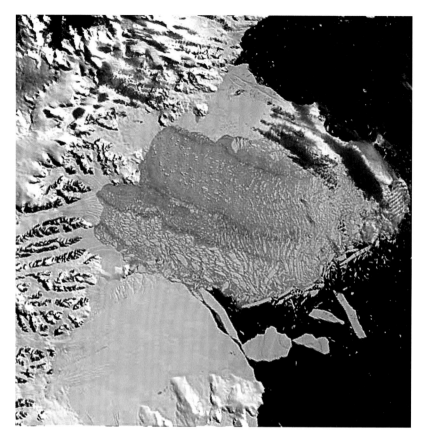

The 2002 fragmentation of Antarctica's vast Larsen B ice shelf (the pale blue area) may speed the thawing of land-based ice, which could raise sea levels. AP Images.

The issue of whether meltwater may lubricate and speed the breakup of ice sheets has been a subject of some debate among scientists. Photographs of meltwater gushing out of moulins, or natural drainpipes, and into the sea have often been used to illustrate global warming's effects. A study led by Roderik S.W. van de Wal of the Institute for Marine and Atmospheric Research suggested that lubrication seems to have little effect over time on loss of ice, however. Instead, meltwater seems to be a seasonal process, speeding up ice movement in summer and slowing it down in winter, but not having a major effect on Greenland ice loss over time. This observation "does not mean that more widespread surface melting could not eventually destabilize vast areas" of Greenland's ice, but it does seem that the impact of moulins in the short term appears to be overestimated, according to environmental reporter Andrew C. Revkin.[16]

Whether lubrication is an important contributor to Greenland ice melt or not, however, there is little doubt that Greenland is losing ice. Van de Wal, the author of the lubrication study, explained that, though positive feedback did not seem to be as catastrophic as some feared, "Greenland will contribute modestly to sea level rise by about 10 cm over this century [just under 4 inches]."[17]

The melting of ice sheets, in other words, does contribute to sea-level rise. For the moment, scientists such as van de Wal believe that its contribution seems modest, though important over time. The systems and mechanisms involved in ice sheet melt are sufficiently complicated, however, that unpredictable dramatic events or changes remain possible.

Notes

1. *Nova: Science in the News*, "Getting into Hot Water—Global Warming and Rising Sea Levels," May 2008. www.science.org.au.
2. Marcel Leroux, *Global Warming: Myth or Reality*. Chichester, UK: Praxis Publishing, 2005.
3. *Earth Observatory*, "Correcting Ocean Cooling," 2007. http://earthobservatory.nasa.gov.
4. *Earth Observatory*, "Correcting Ocean Cooling."
5. *Water Science for School*s, "Glaciers and Icecaps: Storehouses of Freshwater," May 13, 2009. http://ga.water.usgs.gov.
6. Quoted in *Physorg.com*, "Scientists Expect Increased Melting of Mountain Glaciers," January 20, 2006. www.physorg.com.
7. Andrew Weaver, *Keeping Our Cool: Canada in a Warming World*, Toronto, Ontario : Penguin, 2008, p. 177.
8. Quoted in Blake de Pastino, "Mountain Glaciers Melting Faster than Ever, Expert Says," *National Geographic Online*, February 16, 2007. http://news.nationalgeographic.com.
9. *CBC News Online*, "Kilimanjaro Glaciers Could Go Within Decades," November 2, 2009. www.cbc.ca.
10. Quoted in Vince Stricherz, "The Woes of Kilimanjaro: Don't Blame Global Warming," *University of Washington News*, June 11, 2007. http://uswnews.org.
11. Quoted in Blake de Pastino, "Mountain Glaciers Melting Faster than Ever, Expert Says."

12. *UNEP Grid-Arendal*, "Climate Change 2001: Working Group 1: The Scientific Basis," 2003. www.grida.no.
13. David Archer, *The Long Thaw: How Humans Are Changing the Next 100,000 Years of Earth's Climate*. Princeton, NJ: Princeton University Press, 2009, p. 8.
14. Archer, *The Long Thaw*, p. 52.
15. Lester R. Brown, "Analysis: Melting Ice Raises Spectre of Displaced Humanity," *peopleandplanet.net*, June 3, 2009. www.peopleandplanet.net.
16. Andrew C. Revkin, "A Tempered View of Greenland's Gushing Drainpipes," *Dot Earth*, July 3, 2008. http://dotearth.blogs.nytimes.com.
17. Quoted in Andrew C. Revkin, "Greenland Losing Ice, With or Without Lubrication," *Dot Earth*, July 14, 2008. http://dotearth.blogs.nytimes.com.

* * * *

DEBATES

Global ice melt and sea-level rise are events that scientists are still struggling to understand. One of the most unconventional theories is that the increase in temperatures worldwide could result in a new ice age, rather than in overheating.

The logic supporting the concept of a new ice age involves currents. Currents are continuous, directed movements of ocean water. They are caused by a number of factors, including wind and the rotation of the earth. They are also affected by differences in temperature and salinity (or salt content).

The North Atlantic Current

One of the most important currents is the North Atlantic Current, which moves warm surface waters north toward the Arctic. When the current reaches northern latitudes, it gets cooler and denser, and sinks. Then it turns south, pulling deeper cold waters toward the equator. Eventually, in the south, the water warms, rises, and turns back north. The system as a whole essentially functions like a giant, ever-turning conveyor belt.

The North Atlantic Current and the warm water it moves northward are often credited with keeping the weather mild in the north, especially in western Europe. Ruth Curry of the

Woods Hole Oceanographic Institution, for example, noted, "the Atlantic circulation moves heat toward the Arctic, and this helps moderate wintertime temperatures in the high-latitude Northern Hemisphere."[18]

Some scientists have argued that global warming might have an effect on the North Atlantic Current. As Arctic glaciers melt, freshwater may be dumped into the current. As discussed earlier, both temperature and salt-water content have an impact on currents; a giant dump of cold, freshwater could therefore have an effect on how a current flows. The worry is that the current could "rapidly collapse, turning off the huge heat pump and altering the climate over much of the Northern Hemisphere."[19] Some scientists have suggested that the result could be a new ice age.

There does seem to be some possibility that ice-water melt might affect the North Atlantic Current. Science writer John Roach reported that a survey of scientists found that some believed there to be a more than 50 percent chance that the North Atlantic Current might shut down, while others said that there was "no chance" of such a failure.[20]

Even if the North Atlantic Current is affected, however, it is far from clear that the result would be a new ice age. In fact, according to Richard Seager, a climate researcher at Columbia University, Europe's mild climate is not primarily a result of the North Atlantic Current. Instead, in an article in *American Scientist* in August 2006, Seager explained that Europe has mild weather mostly because it has a maritime climate caused by winds blowing off the ocean (maritime climates are discussed more fully in Chapter 1). Seager noted that the climate in Europe, in fact, is similar to the climate in Seattle, a city that is also northerly and mild, but which has no warm currents nearby.

Seager points out that the North Atlantic Current does have some moderating effect on climate. If it were slowed or stopped, he says, it "would have a noticeable but not catastrophic effect on climate. . . . Temperatures will not drop to ice-age levels. . . . The

Water and Ice

Climate Disaster in Popular Culture

Apocalyptic scenarios based upon global warming have been prominent in popular culture. For example, the idea of an ice age caused by global warming served as the basis for the 2004 film *The Day After Tomorrow*. In that film, ice dumped into the North Atlantic Current triggered catastrophically low temperatures, tornadoes, hurricanes, and a giant hailstorm that leveled Tokyo.

An even more prevalent end-of-the-world scenario involves sea-level rise. Thus, as just one example, in the graphic novel *JLA: American Dreams* by writer Grant Morrison and illustrators John Dell and Howard Porter, published in 1997, there is a scene set in the near future that shows the superhero Aquaman swimming past an almost entirely submerged Statue of Liberty. Only the statue's upraised arm breaks the surface; the head is completely underwater. The scene is meant to suggest that global warming has caused the seas to rise and cover the earth.

A submerged Statue of Liberty is a memorable symbol of the dangers of rising sea levels, and it has appeared in both fictional stories, such as comic-book tales, and in actual news stories. For instance, a Boca Raton paper called the *Sun* ran on the cover of its July 25, 1995, issue a picture of water risen to the chest of the Statue of Liberty. Despite their popularity, however, these images have no basis in fact. To immerse the Statue of Liberty to its chest, sea levels would have to rise by about 76m (250 feet), according to University of Victoria climate scientist Andrew Weaver. This is a greater sea level rise "than would occur if all the ice on our planet were to melt," Weaver notes in his book *Keeping Our Cool*.

North Atlantic will not freeze over, and English Channel ferries will not have to plow their way through sea ice." Instead, Seager argues, a North Atlantic slowdown would probably serve merely to provide slight mitigation for human-caused global warming in Europe.[21] In *Keeping Our Cool*, Andrew Weaver observes that "While much has been made in the media of global warming potentially leading to the next ice age, it's simply not possible." Weaver argues in addition that, although a slowdown of the North Atlantic Current might reduce sea-surface temperatures, it probably will not have a significant mitigating effect on global warming in Europe itself.[22]

Changes in currents might not cause a new ice age, but they may nonetheless have other important effects. For example, one research study led by University of Colorado scientists Scott Lehman and Thomas Marchitto suggested that the melting of ice in the Atlantic and resulting changes in currents could result in the ocean releasing stored carbon in a giant burp. Such burps occurred in the past when currents were altered; and future burps might dangerously intensify global warming. "This study provides strong indicators of just how intimately coupled the connection between the ocean and atmosphere can be," said Kent State University professor Joseph Ortiz, a coauthor of the study.[23]

How Much Will Sea Levels Rise?

Scientists believe global warming is very unlikely to cause an ice age. Another dangerous scenario that is often discussed, however, is a massive rise in sea level. In the documentary film *An Inconvenient Truth*, for example, Nobel Peace Prize winner and former U.S. vice-president Al Gore argues that melting ice sheets could cause a sea-level rise of 20 feet (or about 6m) "in the near future."[24] Gore presents a graphic showing inundated coastlines, and contends that this sea-level rise would cause massive dislocations and suffering.

Some scientists agree with Gore that there could be short-term, dangerous increases in sea levels, especially if Greenland's ice sheet were to melt more rapidly than predicted. Jonathan Overpeck, a scientist at the University of Arizona, has used computer models to estimate what would happen if the sea rose by various levels. The models show that a 20 foot (6m) rise would submerge swaths of coastal areas, including, for example, much of Florida. Overpeck noted in a 2004 article, "The consequences would be catastrophic."[25]

In contrast, some scientists argue that global warming may not cause any significant increase in sea levels at all. According to Christopher Essex, a professor of mathematics at the University of Western Ontario, and Ross McKitrick, a professor of economics at the University of Guelph, the effects of thermal expansion of the oceans may be offset by increased evaporation. Essex and McKitrick also argue in their book *Taken by Storm* that reports of actual sea-level variations are inconclusive. One benchmark in Australia, for example, suggests that sea level near Tasmania has actually fallen by some 30cm (about a foot) since 1841. Essex and McKitrick note that "we may know the full explanation behind" the apparent drop in sea level near Tasmania—it might be caused by mistaken measurements, for example. Either way, however, Essex and McKitrick argue that "Even if the sea level does rise, the worst-case forecasts are for something on the order of 50cm [1.5 feet] over . . . 170 years."[26]

> "A commission of 20 international experts" estimates a sea-level rise of 1.5m to 3.5m in sea level by 2200 . . . such an increase "would spell the end of many of our coastal cities."

The mainstream of scientific opinion falls between Al Gore's scenario and the relatively safe future envisaged by Essex and McKitrick. The Intergovernmental Panel on Climate Change (IPCC) Fourth Assessment Report from 2007 estimates that sea levels will rise 18cm to 59cm (7 inches to 23 inches) in the next century. According to a January 2009 report from the U.S.

Following page: *Former U.S. vice-president Al Gore used the documentary film* An Inconvenient Truth *to warn about rising sea level and other effects of climate change.* Valery Hache/AFP/Getty Images.

Climate Change Science Program, however, "Recent observations suggest that sea-level rise rates may already be approaching the higher end of the IPCC estimates.... This is because potentially important meltwater contributions from Greenland and Antarctica were excluded due to limited data and an inability at that time to model ice flow processes adequately. The report concluded that "a global sea-level rise of 1m is plausible within this century if increased melting of ice sheets in Greenland and Antarctica is added to the factors included in the IPCC estimates. Therefore, thoughtful precaution suggests that a global sea-level rise of 1m to the year 2100 should be considered for future planning and policy discussions."[27]

A meter is not the 20 feet (6m) that Al Gore predicted. Still, over a century, it would be a significant increase in sea levels. Moreover, the sea-level rise is expected to continue in future centuries. Stefan Rahmstorf, a climate scientist and oceanographer at the Potsdam Institute for Climate Impact Research, notes that "A commission of 20 international experts" estimates a sea-level rise of 1.5m to 3.5m by 2200. Rahmstorf notes that such an increase "would spell the end of many of our coastal cities."[28]

The disagreements among scientists indicate a frustrating fact about sea levels—they are very difficult to predict. Essex and McKitrick argue that even "Mean sea level is rather tricky to measure." Not only do sea levels vary, but the land itself rises and falls; northern Europe, for example, is rising slightly, while southern Europe is sinking slightly. Computer models attempt to adjust for land shifts, but "there is a lot of uncertainty about how much to adjust the sea level data to account for land movements."[29]

Nor is land movement the only variable. For example, one study led by Julie Loisel of Lehigh University found that peat moss bogs in some regions may be expanding as more water from glaciers reaches them. This change might help offset sea-level rise, as the expanded peat absorbs more water. "When people think of sea level rise, they just look at how much ice is lost and think it all goes right into the ocean.... What we want to stress here is that,

RISE IN SEA LEVELS ON THE ATLANTIC COAST

Station	Rate of Sea-Level Rise (mm per year)	Time Span of Record
Annapolis, MD	3.53 ± 0.13	1928–1999
Atlantic City, NJ	3.98 ± 0.11	1911–1999
Baltimore, MD	3.12 ± 0.16	1902–1999
Boston, MA	2.65 ± 0.1	1921–1999
Charleston, SC	3.28 ± 0.14	1921–1999
Eastport, ME	2.12 ± 0.13	1929–1999
Fernandina Beach, FL	2.04 ± 0.12	1897–1999
Fort Pulaski, GA	3.05 ± 0.20	1935–1999
Hampton Roads, VA	4.42 ± 0.16	1927–1999
Key West, FL	2.27 ± 0.09	1913–1999
Lewes, DE	3.16 ± 0.16	1919–1999
Mayport, FL	2.43 ± 0.18	1928–1999
Miami, FL	2.39 ± 0.22	1931–1999
Montauk, NY	2.58 ± 0.19	1947–1999
New London, CT	2.13 ± 0.15	1938–1999
Newport, RI	2.57 ± 0.11	1930–1999
Philadelphia, PA	2.75 ± 0.12	1900–1999
Portland, ME	1.91 ± 0.09	1912–1999
Portsmouth, VA	3.76 ± 0.23	1935–1999
Providence, RI	1.88 ± 0.17	1938–1999
Sandy Hook, NJ	3.88 ± 0.15	1932–1999
Seavey Island, ME	1.75 ± 0.17	1926–1999
Solomons Island, MD	3.29 ± 0.17	1937–1999
The Battery, NY	2.77 ± 0.05	1905–1999
Washington, DC	3.13 ± 0.21	1931–1999
Willets Point, NY	2.41 ± 0.15	1931–1999
Wilmington, NC	2.22 ± 0.25	1935–1999
Woods Hole, MA	2.59 ± 0.12	1932–1999

Source: James G. Ritus et al., *Coastal Sensitivity to Sea-Level Rise: A Focus on the Mid-Atlantic Region*, U.S. Climate Change Science Program, January 2009, p. 19. www.climatescience.gov.

wait, there's another place it can go," said Loisel.[30] There might be many factors, such as peat bogs, that may reduce or increase sea-level rise. Some may matter only a little; others, such as those that affect the speed of ice melt in Greenland, might have a significant impact.

Notes

18. Quoted in John Roach, "Global Warming May Alter Atlantic Current, Study Says," *National Geographic*, June 27, 2005. http://news.nationalgeographic.com.
19. David Stipp, "The Pentagon's Weather Nightmare: The Climate Could Change Radically and Fast. That Would Be the Mother of All National Security Issues," *Fortune*, February 9, 2004.
20. John Roach, "Global Warming May Alter Atlantic Current, Study Says."
21. Richard Seager, "The Source of Europe's Mild Climate," *American Scientist*, August 2006. www.americanscientist.org.
22. Weaver, *Keeping Our Cool*, p. 147.
23. Quoted in Kent State University, "Study Sheds Light on Earth's CO_2 Cycles, Possible Impacts of Climate Change," May 2007. www.kent.edu.
24. Quoted in Roger Highfield, "*An Inconvenient Truth* Exaggerated Sea Level Rise," *Daily Telegraph*, September 4, 2008. www.telegraph.co.uk.
25. Quoted in Stefan Lovgren, "Warming to Cause Catastrophic Rise in Sea Level?" *National Geographic*, April 26, 2004. http://news.nationalgeographic.com.
26. Christopher Essex and Ross McKitrick, *Taken by Storm: The Troubled Science, Policy, and Politics of Global Warming*, revised edition. Toronto, Ontario: Key Porter Books, 2007, p. 288.
27. James G. Titus et al., *Coastal Sensitivity to Sea-Level Rise: A Focus on the Mid-Atlantic Region*, U.S. Climate Change Science Program, January 2009, pp. 19–20. www.globalchange.gov.
28. Stefan Rahmstorf, "We Must Shake Off This Inertia to Keep Sea Level Rises to a Minimum," *Guardian*, March 3, 2009. www.guardian.co.uk.
29. Essex and McKitrick, *Taken by Storm*, p. 286.
30. Quoted in Michael Reilly, "Sponge-Like Peat Bogs Could Offset Sea Level Rise," *Discovery News*, May 28, 2009. http://dsc.discovery.com.

* * * *

EFFECTS

Despite disagreements about the exact amount, scientists in general do believe that there will be a rise in sea levels due to global warming over the next hundred years. Such a rise will have important effects on coastal areas in the United States.

U.S. Coastlines

Perhaps the chief worry in coastal areas is that sea-level rise will cause a loss of land—especially loss of valuable wetlands. According to the Environmental Protection Agency, "Coastal wetland ecosystems, such as salt marshes and mangroves (trees and shrubs that grow in salt-water habitats) are particularly vulnerable to rising sea level because they are generally within a few feet of sea level. . . . Wetlands provide habitat for many species, play a key role in nutrient uptake, serve as the basis for many communities' economic livelihoods, provide recreational opportunities, and protect local areas from flooding."[31] As sea levels rise, wetlands turn into ocean, and areas further inland become wetland. Researchers think that, in part because of human developments such as dikes and barriers, there may be far fewer wetlands created than there are wetlands destroyed. By 2080, "sea level rise could convert as much as 33 percent of the world's coastal wetlands to open water."[32]

Rising sea levels may also affect drinking water. Cities such as Philadelphia and New York obtain drinking water not far from parts of rivers that are salty during droughts. When sea levels rise, saltwater pushes further upstream and may contaminate water supplies. The water in underground natural aquifers, which store freshwater, may also turn saline. Such contamination could have a major effect on drinking water in the Florida Keys and elsewhere.

Rising seas could also increase the likelihood of damage from storms and flooding. Higher sea levels erode the shores that protect properties from severe weather. Rising seas also cause coastal areas to drain more slowly, because the rate at which water drains depends on the difference in elevation between the place that is draining and the place the water is flowing to. Thus, when sea levels rise, the difference in elevation is reduced, and drainage

CALIFORNIA POPULATION VULNERABLE TO A 100-YEAR FLOOD ALONG THE PACIFIC COAST, BY COUNTY

County	Current Risk	Risk with 1.4m Sea-Level Rise	Percent Increase
Del Norte	1,800	2,600	47%
Humboldt	3,700	7,800	110%
Los Angeles	3,700	14,000	270%
Marin	530	630	20%
Mendocino	530	650	22%
Monterey	11,000	14,000	36%
Orange	72,000	110,000	55%
San Diego	3,000	9,300	210%
San Francisco	4,800	6,500	35%
San Luis Obispo	670	1,300	98%
San Mateo	4,700	5,900	24%
Santa Barbara	3,400	6,700	94%
Santa Cruz	11,000	16,000	49%
Sonoma	580	700	21%
Ventura	7,300	16,000	120%
TOTAL	**130,000**	**210,000**	**67%**

Source: Matthew Heberger et al., *Impacts of Sea-Level Rise on the California Coast*, Oakland, CA: Pacific Institute, 2009, p. 42.

The Wetland Feedback Loop

Rising sea levels are only one of the threats facing wetlands. Humans often drain wetlands to create dry land for agriculture and for homes. Because of such development, in the last century more than 60 percent of wetlands worldwide were destroyed, and more than 90 percent in Europe. In addition, global warming itself has a powerful effect: as temperatures rise, some wetlands evaporate.

The reduction of wetlands in these ways can have a profound impact on global warming. In the first place, wetlands help soak up excess water during storms, and so might help mitigate increased flooding as sea levels rise. Even more important, the destruction of wetlands may itself contribute to climate change. This is because wetlands "hold massive stores of carbon—about 20 percent of all terrestrial carbon stocks," according to Eugene Turner of Louisiana State University. Wetlands cover only 6 percent of the earth's surface, but they hold as much carbon as is currently contained in the earth's entire atmosphere. Wetlands also contain large stores of methane, a greenhouse gas 21 times more potent than carbon. Destroying wetlands, therefore, releases large stores of greenhouse gases, contributing to global warming—which in turn causes temperatures and sea levels to rise, both of which contribute to the destruction of more wetlands.

For all these reasons, many researchers argue that preserving wetlands is a vital part of controlling global warming. After spending so much time and effort draining wetlands, humans now need to figure out how to preserve the remainder, and perhaps even restore some of what has been lost.

---

slows down. Slower drainage, in turn, contributes to flooding. As a result, storm surges from hurricanes, for example, could do serious damage to New York City, which is only 16 feet (5m) above mean sea level. Jianjun Yin, a climate modeler at Florida State University, believes that sea-level rise along the Atlantic Coast would be significantly higher than in other areas because of changes in ocean currents. Yin explains that "The northeast coast of the United States is among the most vulnerable regions to future changes in sea level and ocean circulation, especially when considering its population density and the potential socioeconomic consequences of such changes."[33]

North Carolina is another area that might be especially hard hit, since parts of the coast there are already sinking, potentially compounding rise in sea levels. "Tide-gauge readings in the mid-Atlantic indicate that relative sea level rise [the combination of rising waters and sinking land] was generally higher—by about a foot—than the global average during the 20th century." Some mid-Atlantic barrier islands could end up "destabilizing and breaking apart" as the waters rise. Louisiana, Florida, and other sections of the East Coast would also be threatened.[34]

California, too, is threatened by sea-level rise. The 4-foot (1.4m) increase in sea levels that many scientists expect within the next one or two centuries would "put 480,000 people at risk

A dike near the town of Wilnis, Netherlands collapsed on August 26, 2003, causing the evacuation of 2,000 residents. A rise in sea levels could cause similar problems for many low-lying nations. Continental/AFP/Getty Images.

of a [hundred]-year flood [that is, of a flood expected to occur only once every 100 years] . . . In addition, the cost of replacing property at risk of coastal flooding under this sea-level rise scenario is estimated to be nearly $100 billion."[35]

Global Coastlines

Coastlines in the United States do not face this threat by themselves; Coastlines throughout the world would be changed by rising sea levels. In general, the poorest regions would be the areas hardest hit, whereas wealthier regions would likely be able to adjust more effectively. For instance, the Netherlands is a very low-lying country. Large portions of its land were actually reclaimed from the sea through a system of extensive dikes and dams. These very dikes and dams, however, now provide protection from continued increases in sea levels. This is especially true because the Dutch are relatively wealthy, and they have the resources to expand their water control network. Thus, while "without any doubt . . . the Netherlands will experience potential impacts by rising sea levels," they also have "the technical and financial capacities" to respond to the danger.[36]

If seas were to rise by only 1.5 feet (45cm), which is at the low end of most predictions for the next century, Bangladesh would lose more than 7,500 square miles (about 19,425 sq. km) of land.

In contrast with the Netherlands, Bangladesh is a poor country without many flood control measures. In addition, the land itself is sinking, which adds to the impact of sea-level rise. Thus, if seas were to rise by only 1.5 feet (45cm), which is at the low end of most predictions for the next century, Bangladesh would lose more than 7,500 square miles (19,425 sq. km) of land. If sea level rises by 3.2 feet (1m), Bangladesh could suffer more than $5 billion worth of damage, which is about 10 percent of its gross domestic product (GDP). In addition, rising seas would increase the salinity of the soil, causing crop failures in a nation that already has difficulty in feeding its people. "In view of Bangladesh's already problematic food situation, the expected decrease of rice production, as well as several hundred tons of vegetables, lentils, onions and other crops, could be disastrous," according to Sonja Butzengeiger and Britta Horstmann writing for Germanwatch, a nonprofit nongovernmental public policy organization.[37]

Bangladesh is an extreme case, but poorer nations throughout the world face similar dangers. For instance, in West Africa water may be contaminated and coastlines flooded as seas rise. Nigeria's capital city, Lagos, is only 16 feet (5m) above sea level. It is possible that in the next century a strong tropical storm will cause a surge by which "most of the 15 million inhabitants of Lagos will be displaced."[38]

Islands

Rising sea levels can have a major impact on islands as well as on larger landmasses. Sea-level increase has caused particular concern in the Pacific, where small islands may be entirely engulfed by the ocean if scientist's predictions are accurate. Indeed, according to environmental journalist and former Congressional staffer Curtis A. Moore, some islands, such as Tebua, have already been completely submerged. Moore notes, "Tebua's disappearance is not the only sign that global warming is making itself felt in the distant reaches of the Pacific, where scientists have long predicted that rising waters would engulf low-lying islands. Other islands have disappeared, too. Cemeteries are crumbling into the ocean. Salt has poisoned water supplies. Malaria and other diseases have spread. And large ocean surges have engulfed once-safe homes with no warning." In one dramatic anecdote, from the island of Tarawa, Moore relates how an inhabitant, Teunaia Abeta, "watched as a high tide came rolling in from the atoll's (a coral island circling a lagoon) turquoise lagoon and did not stop. There was no typhoon, no rain, no wind, 'just an eerie rising tide that lapped higher and higher,' according to one account, swallowing up Abeta's thatched-roof home and scores of others."[39]

Even island nations that are not in danger of submersion will face serious challenges. Many of these threats are similar to ones that affect coastal regions: erosion of shorelines, contamination of water supplies, increased flooding during storms. There may be additional problems as well. For example, in Hawaii "standing pools of water will accumulate throughout the region without a place to drain. Travel will be limited and many lands will turn to wetlands."[40]

Although it is clear that rising sea levels will pose a threat to island nations in the Pacific, the exact extent of the threat has been a subject of some debate. One of the most intense discussions has involved Tuvalu, a group of nine atolls with a total land area of only 10 square miles (26 sq. km), located halfway between Hawaii and Australia. In the 1990s, the government of Tuvalu declared its intention to relocate its entire population in order to avoid rising seas caused by global warming. In an article titled "Farewell Tuvalu," on October 29, 2001, reporter Andrew Simms noted that "A group of nine islands, home to 11,000 people, is the first nation to pay the ultimate price for global warming."[41]

It is not at all clear, however, that Tuvalu is actually sinking. In fact, "sea level in Tuvalu has been *falling*—and precipitously so —for decades," according to Patrick J. Michaels, a climatologist and Senior Fellow at the Cato Institute.[42] Michaels suggests that Tuvaluans wanted to evacuate because Tuvalu has few natural resources; the threat of rising waters was simply a convenient excuse. Michaels adds that "Oceans don't rise or fall uniformly around the globe. Instead they primarily respond to local change in ocean temperature and wind."[43] Michaels argues that worries about Tuvalu's submergence are unfounded, and that concerns about sea-level rise on other Pacific island nations may be exaggerated as well.

Other writers have also noted the difficulty of predicting future sea-level rise in the Pacific. For example, the U.S. Global Change Research Program notes that "a number of Pacific islands are themselves rising due to geologic uplift—plate tectonic movements. As a result of this uplift, it can look like sea level is not changing or even falling in some locations. Therefore it is difficult to establish an average relative sea-level rise rate for the Pacific, and the global rise rate can even be hidden."[44]

Mitigating the Damage

Figuring out how to mitigate or prevent the effects of sea-level rise is very difficult to do. As stated earlier, no one is sure by how much sea levels will rise. As discussed above, Bangladesh will experience major problems if the sea rises by only 1.5 feet (45cm) The devastating dislocations and property damage predicted in California, however, are based on a rise of 4 feet (1.5m) or more. To make things even more complicated, different locations will undoubtedly experience different levels of rise depending on local factors, such as rising or falling land. Christopher Essex and Ross McKitrick note that "A 15cm [about 6 inches] change [in sea levels] would require a relatively small investment in dikes and drainage machinery, while a 50cm [about 1.6 feet] change would be much costlier...."[45] On the other hand, the 3-foot (1m) rise suggested by some scientists to occur within a century would require an even more expensive response.

Whatever the actual rise may be, there are some concrete steps that can be taken to reduce the damage from rising sea levels. The U.S. government report *Coastal Sensitivity to Sea-Level Rise* suggests preserving wetlands, building infrastructure where feasible, and adjusting insurance rates in floodplains to reflect increased risks. The report notes "Responding to sea-level rise requires careful consideration regarding whether and how particular areas will be protected with structures, elevated above the tides, relocated landward, or left alone and potentially given up to the rising sea."[46] Replenishing sand on eroding beaches or trying to hold back the sea can keep communities in place, but doing so is expensive. Allowing the sea to swallow land and communities is not an attractive possibility either, however.

In some cases, there seem to be no good options. For example, if seas do rise high enough to threaten Lagos, Nigeria, the capital might have to be abandoned. To move so many people, however, and abandon the nation's economic and political heart would be an "unthinkable option," according to George Awudi, the Ghana programme coordinator for the environmental lobby group Friends of the Earth.[47] Instead, Awudi argues that the best way to combat sea-level rise is for wealthy nations such as the United States to cut greenhouse gas emissions. Unfortunately, even if emissions could be reduced immediately, many scientists believe that the sea level would continue to rise for up to a century. High-end predictions of sea-level rise may well be inaccurate; if they are true, however, it may be too late to avert at least some dire consequences.

Notes

31. *U.S. EPA*, "Coastal Zones and Sea Level Rise," August 3, 2009. www.epa.gov.
32. *U.S. EPA*, "Coastal Zones and Sea Level Rise."
33. Quoted in Jill Elish, "Sea Level Rise Due to Global Warming Poses Threat to New York City," *Florida State University*, March 16, 2009. www.fsu.com.
34. *Guardian*, "New U.S. Report Details Effects of Rising Sea Levels," January 22, 2009. www.guardian.co.uk.
35. Matthew Heberger et al., *Impacts of Sea-Level Rise on the California Coast*. Oakland, CA: Pacific Institute, 2009, p. xi.
36. Sonja Butzengeiger and Britta Horstmann, *Sea-level Rise in Bangladesh and the Netherlands: One Phenomenon, Many Consequences*. Bonn, Germany, 2004, p.5. www.germanwatch.org.
37. Butzengeiger and Horstmann, *Sea-level Rise in Bangladesh and the Netherlands*, p. 6.
38. *IRIN*, "West Africa: Coastline to Be Submerged by 2099," August 25, 2008. www.irinnews.org.
39. Curtis A. Moore, "Awash in a Rising Sea: How Global Warming Is Overwhelming the Islands of the Tropical Pacific," *International Wildlife*, January-February 2002.
40. Chip Fletcher, "The Blue Line," *School of Ocean and Earth Science and Technology at the University of Hawaii at Manoa*, n.d. www.soest.hawaii.edu.
41. Andrew Simms, "Farewell Tuvalu," *Guardian*, October 29, 2001. www.guardian.co.uk.
42. Patrick J. Michaels, *Meltdown: The Predictable Distortion of Global Warming by Scientists, Politicians, and the Media*. Washington, DC: Cato Institute, 2004, p. 204.

Water and Ice

43. Michaels, *Meltdown*, p. 204.
44. *USGCRP*, "Climate Change Impacts on the US: US-Affiliated Islands of the Pacific and Caribbean. Educational Resources," October 12, 2003. www.globalchange.gov.
45. Essex and McKitrick, *Taken by Storm*, p. 280.
46. Titus et al., *Coastal Sensitivity to Sea-Level Rise*, p.6.
47. Quoted in *IRIN*, "West Africa: Coastline to Be Submerged by 2099."

Chapter 4

Rainfall, Hurricanes, and Drought

Warm air holds more water vapor than does cold air. Thus, as the planet warms, there will be more moisture in the air. This will result in more precipitation. "The global increase in rainfall is forecast to be about 3–5%," according to University of Chicago ocean chemist David Archer in his book *The Long Thaw*.[1]

Increased Rainfall

The 3 to 5 percent prediction "seems small," Archer notes, "and perhaps it is." He adds that, for human beings, who need water to live, an increase in rainfall is much better than a decrease.[2] Some researchers suggest that this small rainfall bump could be seriously underestimating the amount by which precipitation may increase, however. A study led by Frank Wentz at the research company Remote Sensing Systems, for example, found that over the past 20 years, as global temperatures have risen, precipitation has increased faster than any scientists expected. Thus, the study suggests that "all models used to predict global warming underestimate the rate at which precipitation increases in response to surface warming," according to University of Miami professor Brian Soden.[3]

Any increase in rainfall, big or small, is going to cause trouble for Bangladesh, where flooding already regularly causes serious damage. Bangladeshis may have gotten a hint of the future in 1998, when "heavy snows and rains in India and Nepal combined

Water and Ice

with increased monsoon downpours and high tides to flood out more than twenty million Bangladeshis. Hundreds were killed."[4]

> *On July 20, 2007, an unprecedented amount of rain fell on England. In parts of Gloucestershire, months of rain fell in only a few hours, rivers overflowed, and water treatment plants were flooded, leaving more than 350,000 people without drinking water.*

Other regions are also threatened by increased precipitation. In 2007, for example, West Africa experienced its heaviest rainfall in decades. Kinshasa, the capital of the Democratic Republic of Congo, was inundated, with 30 deaths caused by the flooding in a single day, while in Ghana 300,000 people had to leave their homes because of the downpour. In all, the deluge "affected 1.5 million people across the continent, killing at least 300 since early summer," according to *National Geographic* reporter Alexis Okeowo.[5]

The title of Okeowo's *National Geographic* article was "Global Warming Causing African Floods, Experts Say." As it suggests, newspapers and other popular sources sometimes directly attribute events like the 1998 Bangladeshi floods or the 2007 African floods to global warming. For example, on July 20, 2007, an unprecedented amount of rain fell on England. In parts of Gloucestershire, months of rain fell in only a few hours, rivers overflowed, and water treatment plants were flooded, leaving more than 350,000 people without drinking water. These dramatic events led Michael McCarthy, the environment editor of *The Independent*, to declare, "It's official: the heavier rainfall in Britain is being caused by climate change, a major new scientific study will reveal this week, as the country reels from summer downpours of unprecedented ferocity."[6]

The study McCarthy referred to was undertaken by Xuebin Zhang and Francis Zwiers of Environment Canada. These scientists, however, did not in fact say "that the July 20 rainfall in

Record Snowfall of 2010 Does Not Contradict Global Warming

We just got five feet of snow in Washington [February 2010] and so everybody is like—a lot of the people who are opponents of [the idea of] climate change, they say, "See, look at that, there's all this snow on the ground." This doesn't mean anything. I want to just be clear that the science of climate change doesn't mean that every place is getting warmer; it means the planet as a whole is getting warmer. But what it may mean is, for example, Vancouver, which is supposed to be getting snow during the Olympics, suddenly is at 55 degrees [Fahrenheit], and Dallas suddenly is getting seven inches of snow.

The idea is, is that as the planet as a whole gets warmer, you start seeing changing weather patterns, and that creates more violent storm systems, more unpredictable weather. So any single place might end up being warmer; another place might end up being a little bit cooler; there might end up being more precipitation in the air, more monsoons, more hurricanes, more tornadoes, more drought in some places, floods in other places. . . .

SOURCE: Barack Obama, Remarks by the President at a Town Hall Meeting in Henderson, Nevada, February 19, 2010.

England was caused by global warming," according to University of Victoria climate scientist Andrew Weaver in his book *Keeping Our Cool*. Instead, Weaver explained, the scientists "provided compelling evidence that the lion's share of the observed increase in middle to high latitude northern hemisphere precipitation . . . was a consequence of human-induced global warming."[7] Thus, scientists cannot tell whether any particular weather event is or is not the result of global warming. But they can tell whether certain weather events will be more or less likely as the weather warms. In this case, scientists did not say that the July 20 floods were caused by global warming. But they did say that serious floods in England were growing more likely because of climate change.

Benefits of Increased Rainfall

Increased rainfall over time would certainly cause a lot of damage in some regions. But it might have some benefits as well. Some climate models have shown dry areas of the earth getting even drier as the world warms. Frank Wentz's study discussed previously, however, suggested not only that rainfall would be higher than expected in wet areas, but that it would be higher than expected in dry areas as well. If this hypothesis is true, drought-plagued regions like the American Southwest might finally get enough water in a warmer world.[8] More generally, increased rainfall might be good for crops. This outcome is especially likely because plants use CO_2 as a nutrient, and because higher temperatures might mean a longer growing season as frosts occur later and thaws earlier. Add together more water, more CO_2, and less frost, and altogether global warming might be "beneficial to agriculture" in many parts of the world, according to David Archer.[9]

More controversial is the suggestion that rainfall might actually help reduce global warming. A few scientists, such as Richard Lindzen of MIT, argue that the increased cloud cover associated with more precipitation may reflect light from the sun, creating a negative feedback loop and reducing the impact of climate change. These arguments, however, "have been widely challenged," according to reporter Alan Prendergast.[10] For instance, James Hansen of the Goddard Institute for Space Studies has argued based on measurements from ice-core samples that past climate changed more quickly than Lindzen's theories suggest. Hansen called Lindzen's theory "simply wrong."[11]

Hurricanes and Extreme Weather Events

As discussed briefly in Chapter 1, one of the most serious worries about rising temperatures is that they will not only increase rainfall and flooding, but that they may contribute to extreme weather events like hurricanes. The primary fear is that rising temperatures may warm the ocean's surface. University of California professor of geography Catherine Gautier, in her book *Oil, Water, and Climate*, explains that "a hurricane is essentially a heat engine with the warm tropical ocean as its primary energy supply. Hurricanes require a sea surface temperature above 28°C [82.4°F] to develop and so are bred in warm tropical regions. . . . Both theory and computer modeling suggest that a warmer ocean (as is expected in a world with a higher concentration of greenhouse gases) could increase the strength of hurricanes because more water vapor is added to the atmosphere by additional evaporation. . . ."[12] More evaporation means more air is rising and mixing, spreading heat and energy throughout the atmosphere—a process known as atmospheric convection. The increase in energy and air movement spurs the development of hurricanes.

The possible link between global warming and hurricanes has led some commentators to attribute Hurricane Katrina, the 2005 hurricane that destroyed New Orleans, to the effects of global warming. Nobel Peace Prize winner and former vice-president Al Gore, for example, suggests in the documentary film *An Inconvenient Truth* that Katrina may have been worsened by climate change. Reporter Jeffrey Kluger also suggests that warmer ocean waters may have contributed to the storm and concludes, "While the people of New Orleans may not see another hurricane for years, the next one they do see could make even Katrina look mild."[13]

In contrast, Christopher Essex, a professor of mathematics at the University of Western Ontario, and Ross McKitrick, a professor of economics at the University of Guelph, argue that Katrina

---

Following page: Hurricane Katrina, which devastated New Orleans in 2005, may have gained greater force from warmer ocean waters. Michael Lewis/National Geographic/Getty Images.

had nothing to do with global warming. Essex and McKitrick point out that hurricanes are not simply a function of ocean temperature; they also require specific atmospheric humidity, specific wind conditions, and so forth.[14] These myriad conditions suggest why hurricanes are so difficult to predict. Furthermore, Essex and McKitrick argue there is no evidence that the number of hurricanes is actually increasing. Although 2005, when Katrina hit land, was a bad year for hurricanes, 2006 was actually a very mild year. In fact, for the United States, "the data show that the 1940s were the worst in terms of both total and major hurricanes, followed by the 1950s, 1930s, and 1890s."[15] In comparison, recent decades have seen relatively mild hurricane activity.

Gautier agrees with Essex and McKitrick that "No sound theoretical grounds yet exist to predict how climate change will affect the number of hurricanes, if at all."[16] She notes, however, that recent satellite data does seem to indicate an increase in the strength, though not the number, of hurricanes. Whether this increase is due to global warming or to natural hurricane cycles is difficult to determine. Gautier concludes that "combined natural and anthropogenic [man-made] effects are likely to induce increased Atlantic hurricane activity in the forthcoming decades" except in years affected by an atmospheric phenomenon called El Niño, which involves the periodic heating of portions of the Pacific Ocean, and which tends to moderate storms.

Other data also raises concerns about the possible effect of temperature rise on storms. For example, researchers at the Carnegie Institution for Science found in a 2008 study that jet streams have risen and shifted toward the poles. Jet streams are high bands of wind that have a major effect on the paths of weather systems and storms. Jet streams tend to weaken storms, so as those air currents move toward the poles and weaken, hurricanes, which usually develop closer to the equator, may become stronger. Researcher Ken Caldeira has stated, "At this point we can't say for sure that this is the result of global warming, but I think it is."[17]

NASA researcher Hartmut Aumann also collected evidence that suggested that warming may increase hurricane activity. Aumann and his team tracked the formation of high clouds associated with severe storms over the course of the year. They found that when sea surface temperature increased, so did the formation of clouds. They concluded that if temperatures increased because of global warming by .23°F (.13°C) per decade, there was likely to be an increase in storm activity of 6 percent per decade. Aumann has been careful to note the difficulty of making predictions in this area, however. "Clouds and rain have been the weakest link in climate prediction," he said. "The interaction between the daytime warming of the sea surface under clear-sky conditions and increases in the formation of low clouds, high clouds, and ultimately rain is very complicated."[18] Researchers, then, do generally believe that increases in ocean temperatures affect hurricanes. There is still much work to be done, however, before scientists can determine whether, or to what degree, hurricanes are likely to worsen.

Hotter, Drier Climates

In general, scientists expect the world to become wetter as it grows warmer, with more rainfall, more severe storms, more flooding, and possibly more hurricanes. Higher temperatures may also cause drought and dryness in some areas, however. David Archer notes that there is a particular danger of drier climates in continental interiors, which could threaten major agricultural regions such as the Great Plains of the United States. Archer also says that "The greenhouse climate has the potential to produce what are called mega-droughts, lasting for a decade or longer."[19]

An example of one region that could be hard hit by mega-droughts is West Africa. A group of researchers led by Jonathan Overpeck of the University of Arizona recently studied lake sediments in the region. They discovered that the climate in West Africa used to be much drier than it is today. Over the last 3,000

PROJECTED CHANGES IN PRECIPITATION FROM 1980–99 TO 2090–99

December–February

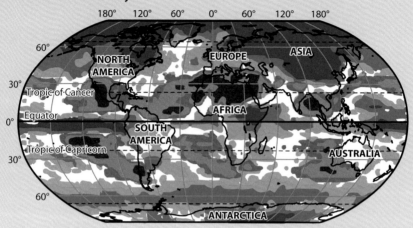

June–August

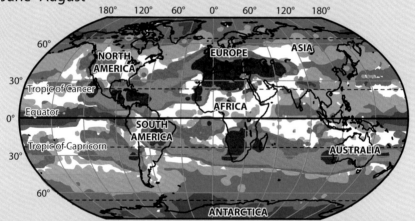

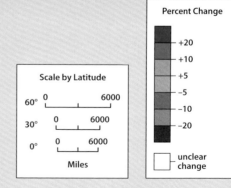

Source: *Encyclopedia Britannica Online*, "Projected Changes in Precipitation from 1980–99 to 2090–99," 2008.

Global Warming May Have Contributed to Slaughter in Darfur

Darfur, a region of western Sudan, is one of the most troubled and violent areas in the world. Since 2003, it has been the site of a bitter civil war between Arab militias (possibly allied with the Sudanese government) and black rebel groups. The war has been characterized by mass murders and rapes of civilians, and it is often referred to as a genocide.

In general, the fight in Darfur is seen as an ethnic conflict. In a 2007 article in the *Washington Post*, however, Ban Ki Moon, secretary-general of the United Nations, argued that the war had begun "as an ecological crisis, arising at least in part from climate change."

Moon noted that some scientists believe that the rise in Indian Ocean temperatures associated with global warming may have been responsible for drought conditions afflicting Sudan. The droughts, in turn, caused a breakdown in relations between nomadic Arab groups and settled black farmers. Before the drought, both groups had shared the same wells and camels had grazed freely. But as the drought intensified, the farmers fenced their lands, fearing that grazing would damage their crops and deplete their water. It was this conflict over resources, Moon argues, that precipitated the civil war.

If Moon is right, Darfur demonstrates graphically two major points about climate change. First, the effects of global warming are extremely hard to predict. And second, whatever those effects are, they will hit those regions with the least resources hardest. Places like the United States, with more resources and stable governments, have some room to maneuver in adjusting to global warming. In places like Sudan, however, a change in climate can spiral quickly into a political and humanitarian catastrophe.

years, the region frequently experienced droughts that lasted decades or even centuries. The researchers also discovered that fluctuations in dryness over the centuries seemed to be related to variations in sea surface temperature. There is, therefore, concern that global warming could tip the region back into a period of extreme drought. In a 2009 article, Overpeck is quoted as saying, "Clearly, much of West Africa is already on the edge of sustainability . . . and the situation could become much more dire in the future with increased global warming."[20]

Another region that may already be experiencing a shift to a drier climate is the American Southwest. There is "broad consensus amongst climate models that this region will dry significantly in the twenty-first century, and that the transition to a more arid climate are already under way. If these models are correct . . . the levels of aridity in the recent multi-year drought, or the Dust Bowl and the 1950s droughts, will, within the coming years to decades, become the new climatology of the American Southwest," according to Columbia University researcher Richard Seager. Seager noted that the United States is wealthy enough that the drought may not cause massive loss of life the way that comparable dry spells have in Africa. He noted, however, that in Western nations irrigation systems and pumps are used widely to bring water vast distances and allow people to live in dry regions that would normally be uninhabitable. Seager warns, "Plumbing on a continental scale supports massive agricultural, industrial and cultural production. Just how vulnerable is such a complex, water-dependent society in an arid region to climate? We do not know. . . . But, as man changes the climate, we may be about to find out."[21]

There are numerous ways in which higher temperatures may contribute to drought. The first is by reducing precipitation. As mentioned earlier, high temperatures usually mean more rainfall. The effect can vary regionally, however. For example, environmental photojournalist Gary Braasch, in his book *Earth Under Fire*, notes that lower rainfall in southern Africa from the late 1970s to today may be related to "higher sea surface temperatures in the Indian Ocean," perhaps caused by global warming.[22] Catherine Gautier notes that models also "predict a widespread decrease in mid-latitude summer precipitation except in eastern Asia, as well as a decrease in precipitation over many subtropical areas."[23]

Even if rainfall does not decrease, however, some regions may still become drier. This is because of evaporation. As temperatures rise, water turns to water vapor more quickly. "The higher evaporation rates will lead to greater drying of soils and vegetation, especially during the warm season."[24] Thus, for example, evaporation rates contributed to the 2002 drought in Australia. A study by the World Wildlife Fund-Australia examined both the 2002 drought and four other droughts since 1950 and concluded that "higher temperatures caused a marked increase in evaporation rates from soil, watercourses, and vegetation."[25] Because there was more evaporation, the drought was more severe, creating a greater danger of brushfires and reducing agricultural productivity.

Increased evaporation rates could have a profound effect on such climates as the Amazon rain forest. "Ultimately, the rain forests in Amazonia are there because the rainy season is so dependable from year to year, and average temperatures, though hot, are not hot enough to evaporate critical amounts of water out of soaked soils," according to University of California-Berkeley paleontologist Anthony D. Barnosky in his book *Heatstroke*.[26] Barnosky looks specifically at the rainforest in Tambopata, Peru, which receives 80 to 100 inches (2,000–2,500mm) of rain per year, so that the soil is basically always wet. If temperatures rise, however, evaporation will happen more quickly, and the ground may become drier. "Dry those soils enough," Barnosky says, "and what used to be rainforest becomes savannah [a tropical grassland with only scattered trees]."[27]

Thus, because of increased evaporation, climates may become drier in certain regions even if rainfall stays the same. By the same token, rainfall increases in some regions may create wetter climates despite higher temperatures and increased evaporation. Because so many complicated factors are involved—rainfall, ocean temperature, cloud cover, evaporation—no one can say for certain that any particular change in a climate is caused by global warming. But scientists do know that, as the earth warms, patterns of precipitation and evaporation will change in ways that will affect everyone, from Bangladesh to Peru.

Notes

1. David Archer, *The Long Thaw: How Humans Are Changing the Next 100,000 Years of Earth's Climate*. Princeton, NJ: Princeton University Press, 2009, p. 47.
2. Archer, *The Long Thaw*, p. 47.
3. Quoted in Anne Minard, "Global Warming Models Underpredict Increase in Rainfall, Study Says," National Geographic, May 31, 2007. http://news.nationalgeographic.com.
4. Gary Braasch, *Earth Under Fire: How Global Warming Is Changing the World*. Berkeley, CA: University of California Press, 2007, p. 121.
5. Alexis Okeowo, "Global Warming Causes African Floods, Experts Say," *Independent*, October 29, 2007. http://news.nationalgeographic.com.
6. Michael McCarthy, "England Under Water: Scientists Confirm Global Warming Link to Increased Rain," *Independent*, July 23, 2007. www.independent.co.uk.
7. Andrew Weaver, *Keeping Our Cool: Canada in a Warming World*. Toronto, Ontario: Penguin, 2008, p. 13.
8. Minard, "Global Warming Models Underpredict Increase in Rainfall, Study Says."
9. Archer, *The Long Thaw*, p.54.
10. Alan Prendergast, "The Skeptic," *Denver Westword News*, June 29, 2006. www.westword.com.
11. Quoted in Fred Guterl, "The Truth About Global Warming," *Newsweek*, July 23, 2001. www.newsweek.com.
12. Catherine Gautier, *Oil, Water, and Climate: An Introduction*. New York: Cambridge University Press, 2008, p. 291.
13. Jeffrey Kluger, "Is Global Warming Fueling Katrina?" *Time*, August 29, 2005. www.time.com.
14. Christopher Essex and Ross McKitrick, *Taken by Storm: The Troubled Science, Policy, and Politics of Global Warming*, revised edition. Toronto, Ontario: Key Porter Books, 2007, p. 279.
15. Essex and McKitrick, *Taken by Storm*, p. 279.
16. Gautier, *Oil, Water, and Climate*, p. 291.
17. Quoted in Carnegie Institution for Science, "Changing Jet Streams May Alter Paths of Storms and Hurricanes," April 17, 2008. www.ciw.edu.
18. Quoted in *USA Today*, "NASA: Global Warming to Increase Severe Storms, Rainfall," December 19, 2008. www.usatoday.com.
19. Archer, *The Long Thaw*, p. 47.
20. Quoted in National Science Foundation, "West African Droughts Are the Norm, Not an Anomaly," April 16, 2009. www.nsf.gov.
21. Quoted in Allen Best, "Thirsting for Water," *Forest Magazine*, Summer 2007.
22. Braasch, *Earth Under Fire*, p. 136.
23. Gautier, *Oil, Water, and Climate*, p. 183.

24. Union of Concerned Scientists, "Early Warning Signs of Global Warming: Droughts and Fires," November 10, 2003. www.ucsusa.org.
25. WWF-Australia, "New Report Shows Global Warming Link to Australia's Worst Drought," January 14, 2003. www.wwf.org.au.
26. Anthony D. Barnosky, *Heatstroke: Nature in an Age of Global Warming*. Washington, DC: Island Press, 2009, p. 138.
27. Barnosky, *Heatstroke*, p. 139.

Chapter 5

Water Supply

Global warming will alter precipitation patterns and evaporation rates. As a result of such changes, the amount of water available for human use in many regions may decrease.

Rivers and Water Supply

Potentially, one of the most dangerous effects of global warming is a decrease in river flow. According to a study published in the *Journal of Climate*, rivers worldwide are losing water as a result of global warming. The study, undertaken by scientists from the National Center for Atmospheric Research (NCAR), looked at river flow from 925 rivers between 1948 and 2004. They found that stream flow decreases were more than twice as common as increases. Decreases were especially common in populated areas: among the rivers losing water were the Yellow River in China, the Niger in West Africa, the Colorado in the United States, and the Ganges in India. All in all, river water flowing into the Pacific Ocean fell by 6 percent, or by as much water as flows into the ocean from the Mississippi every year.

Some of the reduction in river flow in these areas was due to dam construction and the diversion of water for irrigation. The Columbia River, for example, lost 14 percent of its flow over the 50 years of the study, partially because of increased water usage. The reduced flow also reflected a fall in the region's precipitation, however, which could be linked to climate change. Worldwide,

"The researchers found . . . that the reduced flows in many cases appear to be related to global climate change, which is altering precipitation patterns and increasing the rate of evaporation."[1]

Some rivers did show increased flows. The Mississippi, for example, increased by 22 percent because of greater rainfall in the Midwest. Rivers in sparsely populated areas in the Arctic, where increased melting has dumped freshwater into rivers, also increased their flows. So did the Yangtze River, though scientists worry that this river may be threatened in the future as the Himalayan glaciers that feed it melt away.

Falling river flows are worrisome not only because of the dangers to human water supplies, but also because freshwater has an important impact on the oceans. Rivers deposit nutrients and minerals into the oceans, and these nutrients are vital for many sea creatures. In addition, river discharge affects ocean circulation and currents. So far, the changes in river flow have not been large enough to cause major changes in ocean patterns, but researcher Aiguo Dai has noted that "freshwater balance in the global oceans needs to be monitored for any long-term changes."[2]

Global climate change may affect river water in other ways. As mentioned in Chapter 3, when sea levels rise, salt water moves further upstream in rivers. This encroachment can contaminate some water supplies, especially in such places as New York City, Philadelphia, and California's Central Valley, which have water intakes only "slightly upstream from the point where water is salty during droughts."[3] Similarly, high evaporation rates and changing precipitation patterns can result in contaminated freshwater supplies. In the U.S. drought in the summer of 1999, for example, there was not enough freshwater to cleanse rivers and streams; as a result, "salt water encroached further up rivers in many areas of the mid-Atlantic coast."[4]

Lakes and Water Supply

Like rivers, lakes may experience dwindling water supplies as the world warms. In the United States, for example, scientists have been concerned about the shrinking of the Great Lakes. In 2007, Lake Superior's water level was more than 1.5 feet (45cm) below its long-term mean. In part this drop was because of a drought, but the loss of water also seems to be due to long-term trends. Lake surface temperatures have risen by 4.5°F (2.5°C) since 1979, causing an increase in the rate of evaporation. Specifically, evaporation has increased by 1.8 inches (4.6mm) per year since 1978, while precipitation has fallen by .16 inches (4.1mm).[5]

The temperature rise on the lake is greater than air temperature rise over the same period. University of Minnesota researcher Jay Austin has connected the change in lake temperature to "a reduction in winter ice cover on the lake."[6] Ice reflects sunlight; without the ice cover, the lake absorbs more energy from the sun, and water temperatures rise. Ice also puts a cap on the lake that prevents evaporation. With less ice, evaporation increases. Thus, there is a feedback effect: As global warming causes air temperature to rise, there is less ice on the lake, which causes water temperature to rise further.

Since 2007, the Midwest has experienced increased rainfall and colder weather. As a result, according to reporter Sheri McWhirter, "Water levels in the Great Lakes are rising after receding for a decade."[7] Lake Huron and Lake Michigan were 10 inches higher in June than the previous year. Despite such short term gains, however, McWhirter also noted that average lake levels are still expected to decline by about 3 feet (1m) over the course of the century. University of Michigan researcher Don Scavia explains that "Climate projections say the lakes will go up and down around a decreasing average.... The lows will be lower than in the past, and the highs will be lower than in the past."[8]

Less water in the Great Lakes can be problematic. When lake levels were at historic lows in 2007, merchant ships had to carry less cargo or risk being beached. In places like Africa, however, the effect of global warming on lakes has raised the specter of environmental devastation. In Lake Tanganyika in eastern Africa, for example, air temperatures above the lake rose by 1.1°F

Water Supply

Lake Chad Disappears

Lake Chad sits on the borders of Chad, Niger, Nigeria, and Cameroon in West Africa. It has long been an important source of fish and of water for irrigation.

Soon, however, the people of the region may no longer be able to rely on Lake Chad. That is because the lake is disappearing. Since 1963, the surface area of Lake Chad has decreased from 9,562 square miles (25,000 sq. km) in 1963, to just 521 square miles (1,350 sq. km) in 2007.

The dwindling of the lake has been catastrophic for those who live near it. Lake Chad used to yield 253,452 tons (230,000 metric tonnes) of fish; now that has dropped to only 55,116 tons (50,000 metric tonnes.) There is not enough water to irrigate or moisten the soil, and there have been conflicts between herders and farmers over land rights. The people of eastern Nigeria are especially threatened; the lake has run away from them, across the border with Chad. When they try to follow it to claim a portion of the few remaining fish, they are frequently harassed by officials.

Scientists believe that up to 50 percent of the lake loss may be caused by an unsustainable increase in irrigation, which multiplied by four times between 1983 and 1994. Another factor contributing to the loss of water has been overgrazing, which has deforested the region, creating an overall drier climate. And in general throughout Africa, global warming has played a part in the loss of water supplies, according to a study by NASA and the German Aerospace Centre.

As a result of all of these factors, Lake Chad has shrunk to almost a puddle. Muhammadu Bello, a fisherman from the region, told BBC reporter Senan Murray in 2007, "Some 27 years ago when I started fishing on the lake, we used to catch fish as large as a man." No longer.

($.6\,°C$), according to a 2003 study by Catherine O'Reilly of Vassar College, Andrew Cohen of the University of Arizona, and other researchers. The change in air temperature above Tanganyika decreased water circulation, which prevented nutrients' rise from deep waters to surface waters. As nutrient levels fell, algae at the lake's surface also decreased. This decline, in turn, meant that there was less food for fish.

Declining fish stocks are a serious problem, because people around Tanganyika eat 200,000 tons of fish from the lake every year. That amount is about 25 percent to 40 percent of the protein consumed by the population in the region. The scientists predicted that with more global warming stocks might decline by as much as 30 percent. Researcher Andrew Cohen noted, "Given the already significant problems of malnutrition and civil conflict in central Africa, a significant decline in fishing yields resulting from climate change could lead to extremely serious consequences for the region's food supply."[9]

Melting Snow and Water Supply

Both lakes and rivers are affected by melting snow, which is an important source of freshwater for people throughout the world. Changes in snowmelt rate could have a serious effect on the availability of water.

> "Between 1950 and the 1990s, even though overall precipitation rose slightly, the volume of Rocky Mountain snowpack declined by 16 percent."

Although changes in precipitation patterns are uncertain and difficult to predict, scientists are relatively certain that there will be less snow as the temperature warms. In particular, global warming may affect snowpack—naturally formed, compressed snow, especially in mountainous regions, that often melts in springtime. Thus, University of California professor of geography Catherine Gautier, in her book *Oil, Water, and Climate*, notes that "Perhaps the most reliable prediction of change" in the distribution of water "is the decrease in springtime snowpack over the Northern Hemisphere."[10] In 2007, Greg Nickels, the mayor of Seattle, wrote that snowpack in the Cascade Mountains in the Pacific Northwest had declined by half since 1950 "and will be cut in half again in 30 years" without action to address climate change.[11]

The decrease in water from snowmelt is caused by two factors. First, as temperatures warm, precipitation is more likely to fall as rain than snow. Thus, less snow lands on the ground, and there is less of a buildup in snowpack. Second, snow melts earlier as the world warms. This can result in problems for irrigation systems and reservoirs that depend on regular and predictable snowmelt to replenish water supplies.

In his book *Dead Pool*, James Lawrence Powell, executive director of the National Physical Sciences Consortium at the University of Southern California, explains how changes in snow patterns affected the Colorado Basin. "Even a slight reduction in snowpack spells trouble. So much of the water in the Colorado River basin evaporates that nearly 90 percent of the water in streams must come from a virtual reservoir: the Rocky Mountain snowfields. Rising temperatures cause more winter precipitation to fall as rain, reducing snowpack levels at the outset. These reductions have already begun. Between 1950 and the 1990s, even though overall precipitation rose slightly, the volume of Rocky Mountain snowpack declined by 16 percent."[12] Powell goes on to explain how snow melting early can have an adverse effect on water supplies. "The higher the temperature, the earlier the snowpacks start to melt. . . . Earlier melting sends water downstream in the spring, before cities and fields can use it. If reservoirs have room, they can store the earlier-arriving water; if not, they will overflow."[13]

The combination of less snowfall and earlier snowmelt may result in crippling droughts in the western United States, according to Tim Barnett of the University of California, San Diego. Barnett used computer models to analyze the water flows in Western rivers over 50 years. He found that changes in western snowmelt "have less than a one percent chance of being due to

PREDICTED CHANGE IN DISCHARGE OF SELECTED RIVER MOUTHS DUE TO CLIMATE AND WATER USE CHANGES

Continent	River Mouth	Discharge ($km^3\ yr^{-1}$)		Relative Change (%)
		1960s	2050s	
Africa	Kouilou	28.4	20.0	−29.6
	Cross	59.9	61.8	3.1
	Chari	29.1	34.3	17.9
	Senegal	5.7	2.5	−56.0
	Congo (Zaire)	1,349.0	1,267.5	−6.0
	Volta	32.8	48.1	46.7
Asia	Cá	22.3	20.8	−6.7
	Chu Salween (Thanlwin)	98.5	135.2	37.2
	Nadym	16.0	26.5	65.9
	Kura	22.0	13.7	−37.8
	Ganges–Brahmaputra	1,186.9	1,388.4	17.0
	Indus	121.2	174.6	44.1
Australasia	Merauke	1.5	0.9	−40.6
	Fly	135.4	147.5	9.0
	Sepik	100.7	133.6	32.7
	Murray	11.1	9.5	−14.3
	Ramu	32.7	40.1	22.8
Europe	Adour	6.5	6.2	−4.9
	Pechora	142.0	174.1	22.6
	Mezen	26.5	32.8	23.8
	Kuban	13.0	9.7	−25.1
	Volga	234.0	246.3	5.2
	Severn, Dvina	101.2	123.4	21.9
North and Central America	Patuca	12.3	3.4	−72.0
	Yukon	187.2	246.0	31.4
	Kobuk	0.2	0.8	212.2
	Grande de Matagalpa	30.1	7.7	−74.3
	Mississippi	530.6	540.0	1.8
	Colorado	1.3	2.4	81.6
South America	Coppename	10.7	0.7	−93.4
	Essequibo	155.1	78.8	−49.2
	Santa Cruz	0.9	1.0	17.3
	Parnaiba	26.6	5.0	−81.2
	Amazonas–Orinoco	6,802.4	5,536.5	−18.6
	Doce	24.4	33.4	37.1

Source: Margaret A. Palmer et al., "Climate Change and the World's River Basins: Anticipating Management Options," *Frontiers in Ecology and the Environment*, vol. 6, no. 2, 2008, pp. 81–89.

natural variability." Barnett suggested that more water rationing and more dam building might be necessary to avert a crisis. "Global warming is an abstraction to most people," Barnett said. "Well, the people who live in the West, if they haven't already, are going to very shortly find out what global warming really means to them."[14]

Changes in snowmelt will affect people far beyond the American West as well. According to *Climate Change 2007*, part of the Fourth Assessment Report of the Intergovernmental Panel on Climate Change, regions that rely on Himalayan ice melt will see a reduction in water supply. The report also notes that "In the Andes, glacial melt water supports river flow and water supply for tens of millions of people during the long dry season." Ultimately, the report says, "With more than one-sixth of the earth's population relying on melt water from glaciers and seasonal snow packs for their water supply, the consequences of projected changes for future water availability, predicted with high confidence and already diagnosed in some regions, will be adverse and severe."[15]

Groundwater and Water Supply

Changes in snowmelt may also affect groundwater. Groundwater is water located underneath the ground, either in porous soil or in spaces in rock. An underground area that can yield usable water is called an aquifer. Groundwater from aquifers is an important source of water for both agriculture and industry.

Researchers believe that snowfall from winter storms is the major source for replenishing groundwater in the southwestern United States. This is because groundwater recharge tends to happen during floods such as those that occur with winter melt. Thus, "As the snowline retreats to cover smaller and smaller areas, and as the snowpack itself declines because of more rain and less snow and more intermittent melting . . . it seems really likely that recharge will decline in many parts of the Southwest," according to U.S. Geological Survey researcher Michael Dettinger.[16]

Groundwater is an especially important source for human and agricultural use in Africa. With climate change altering rainfall, the flow of rivers, and the level of lakes, groundwater is expected to be both more heavily relied upon and a less reliable source of drinking water on the African continent. Despite its importance in Africa and elsewhere, however, "The impact of climate variability and change on groundwater resources remains . . . one of the most persistent knowledge gaps identified by the Intergovernmental Panel on Climate Change. . . . " according to geographer Richard Taylor.[17]

Water for Crops

Loss of water from rivers, lakes, snowmelt, and groundwater means a reduction in drinking water. But it also means a reduction in water for crops.

As discussed in Chapter 4, global warming may cause an increase in droughts in some areas. Even without accounting for such droughts, however, global warming could have a serious effect on world agriculture. This is because hotter weather often means lower crop yields. A study by David Battisti of the University of Washington and Rosamond Naylor of Stanford University suggested that higher heat in Europe could affect food; in 2003, for example, when Europe saw record temperatures, wheat production in France and Italy dropped by a third. Higher temperatures in the tropics could also cut crop yields by 20 percent to 40 percent, according to Battisti and Naylor. Because the tropics is "home to about half the world's population, the human consequences of global climate change could be enormous."[18]

Scientists are also concerned that higher temperatures will increase the water needed for irrigation, even as water supplies in general are decreasing. According to the Intergovernmental Panel on Climate Change (IPCC), "Higher temperatures and

---

Following page: California's lengthy drought in the early twenty-first century emptied reservoirs and left irrigation canals dry. David McNew/Getty Images.

increased variability of precipitation would, in general, lead to increased irrigation water demand even if the total precipitation during the growing season remains the same." Increased evaporation at higher temperatures would mean that crops would need more water; irregular rainfall would also require more irrigation to make sure that crops had a steady water supply. Thus, even if rainfall stays the same, more irrigation may be needed for crops. The paper notes that irrigation requirements for China and India, the countries with the largest irrigated areas on the earth, might change by 2 percent to 15 percent for China and by −6 percent to +5 percent in the case of India.[19]

Some scientists have argued that worries about global crop shortages are exaggerated. In the first place, as noted in Chapter 4, the combination of increased overall rainfall, higher levels of carbon dioxide, and longer growing seasons may actually increase crop yields in many places. Patrick J. Michaels, a climatologist and Senior Fellow at the Cato Institute, asks, "If the defining characteristics of greenhouse warming are warmer winters, more rain, and longer growing seasons, what's so bad about that?"[20]

Some commentators have also argued that the threats of world crop disaster fail to take into account human adaptation. They maintain that when the world warms, farmers will not just plant the same crops in order to watch them wither in the field. Instead, farmers will adjust, changing what they plant to reflect the new climate. In some cases, the change in climate may even allow farmers to switch to more productive crops. "If the climate changes, you may be worse off if you don't adapt, but you may be better off than you were before if you do," according to Christopher Essex, a professor of mathematics at the University of Western Ontario, and Ross McKitrick, a professor of economics at the University of Guelph.[21]

One way in which humans might adapt to climate change is by creating new crops. For example, Brazil is worried that a rise in global temperatures could "mean a 10% reduction in Brazil's arable land for coffee by 2020." To meet this threat, Brazilian scientists are working to develop genetically modified crops that will be able to withstand the coming heat. Such crops take many years to develop, but Brazil has already seen some success with modified soy plants "that respond favorably to dry, hot conditions while thriving in normal weather as well."[22] Experiments with coffee have been less successful, however.

Despite the hopeful signs, however, the short-term and medium-term outlook is worrisome, according to a 2008 article by *U.S. News & World Report* writer Kent Garber. Discussing a U.S. government report on the impact of global warming on crops, Garber notes that, in the United States, "Some crop yields are predicted to drop; growing seasons will get longer and use more water; weeds and shrubs will grow faster and spread into new territory, some of it arable farmland; and insect and crop disease outbreaks will become more frequent."[23] Garber notes that government action might mitigate problems in the long term, but over the next 30 years at least, America and the world will have to face the fact that higher temperatures may reduce crop yields.

Notes

1. University Corporation for Atmospheric Research, "Water Levels Dropping in Some Major Rivers as Global Climate Changes," April 21, 2009. www.ucar.edu.
2. Quoted in University Corporation for Atmospheric Research.
3. *U.S. EPA*, "Coastal Zones and Sea Level Rise," August 3, 2009. www.epa.gov.
4. Union of Concerned Scientists, "Early Warning Signs of Global Warming: Droughts and Fires," November 10, 2003. www.ucsusa.org.
5. Jessica Marshall, "Global Warming Is Shrinking the Great Lakes," *New Scientist*, May 30, 2007. www.newscientist.com.
6. Marshall, "Global Warming Is Shrinking the Great Lakes."
7. Sheri McWhirter, "Great Lakes Water Levels Rise After Decade of Drops," *Detroit Free Press*, July 26, 2009. www.freep.com.
8. Quoted in McWhirter, "Great Lakes Water Levels Rise After Decade of Drops."
9. Quoted in *National Science Foundation*, "Lake Ecosystem Critical to East African Food Supply Is Threatened by Climate Change," August 13, 2003. www.nsf.gov.
10. Catherine Gautier, *Oil, Water, and Climate: An Introduction*. New York, NY: Cambridge University Press, 2008, p. 185.
11. Greg Nickels, "State Should Find Way to Protect City Light's Climate Protection Efforts," *Seattle Times*, February 7, 2007. http://community.seattletimes.nwsource.com.
12. James Lawrence Powell, *Dead Pool: Lake Powell, Global Warming, and the Future of Water in the West*. Berkeley, CA: University of California Press, 2008, p. 177.
13. Powell, *Dead Pool*, p. 177.

Water and Ice

14. Quoted in Richard A. Lovett, "Western U.S. Faces Drought Crisis, Warming Study Says," *National Geographic Online*, January 31, 2008. http://news.nationalgeographic.com.
15. Martin L. Parry et al., *Climate Change 2007: Impacts, Adaptation and Vulnerability. Contribution of Working Group II to the Fourth Report of the Intergovernmental Panel on Climate Change*. Cambridge, UK: Cambridge University Press, 2007, p. 187.
16. Quoted in Melanie Lenart, "Global Warming Could Affect Groundwater Recharge," *Southwest Climate Outlook*, November 2006. www.climas.arizona.edu.
17. Quoted in University College London, "International Conference on Groundwater and Climate in Africa," August 13, 2008. www.ucl.ac.uk.
18. Quoted in Doyle Rice, "Study Warns of Dire Overheating of Crops, Food Crisis by 2010," *USA Today*, January 13, 2009. www.usatoday.com.
19. Intergovernmental Panel on Climate Change, *Climate Change and Water: IPCC Technical Paper VI*, June 2008, p. 44. www.ipcc.ch.
20. Patrick J. Michaels, *Meltdown: The Predictable Distortion of Global Warming by Scientists, Politicians, and the Media*. Washington, DC: Cato Institute, 2004, p. 163.
21. Christopher Essex and Ross McKitrick, *Taken by Storm: The Troubled Science, Policy, and Politics of Global Warming*, revised edition. Toronto, Ontario: Key Porter Books, 2007, p. 271.
22. Marco Sibaja, "Climate Change Threatens Brazil's Coffee Crop," *USA Today*, February 19, 2009. www.usatoday.com.
23. Kent Garber, "How Global Warming Will Hurt Crops," *U.S. News & World Report*, May 28, 2008. www.usnews.com.

CHAPTER 6

Water and Energy

Scientists assert that climate change is driven by the burning of fossil fuels, which emit greenhouse gases such as carbon dioxide. The National Resources Defense Council (NRDC), an environmental group, notes that "Coal-burning power plants are the largest U.S. source of carbon dioxide pollution." One of the main suggestions scientists have proposed for reducing fossil fuel, therefore, is to move away from coal and petroleum and toward renewable energy resources: As the NRDC says, "We can increase our reliance on renewable energy sources such as wind, sun and geothermal."[1]

Another potential source of renewable energy is water power, or hydropower. Water power comes in a number of different forms.

Hydropower

Water wheels are perhaps the most straightforward example of hydropower, and also the oldest: They were used in ancient Greece and Rome, in ancient China, and throughout the Middle Ages. A water wheel is basically what it sounds like; a large wooden or metal wheel in a river or stream. Blades or buckets are placed on the outside rim of the wheel. These blades or buckets catch the flowing water, which causes the wheel to turn. The wheel is hooked up by gears and belts to machinery, and so can provide power for various activities. Water wheels were most often used to grind flour in mills, but they were also historically used in mining, and for such tasks as crushing wood or cutting marble.

Tidal Power

Today hydropower usually is used to generate electricity rather than to generate mechanical energy directly, as with historical water wheels. One technology in contemporary use for doing this is tidal power. Tidal power harnesses the energy resulting from the ebb and flow of water that occurs as tides come to shore and retreat from it.

The Severn estuary in Wales has a very large difference between high and low tides, and as a result a barrier across it could generate as much energy as eight coal plants.

Tidal energy can be captured in a couple of ways. First, some technologies make use of estuaries. An estuary is a kind of coastal lake into which a river flows. Instead of being entirely enclosed like a lake, however, one part of the estuary is open to the ocean. This means that tides flow into and out of the estuary. To capture the power of the tides, engineers build tidal barrages across the estuary opening. These often look like, and work in a manner similar to, wind turbines. The tidal turbines have blades that can rotate, somewhat like those on a fan. Water flows into the estuary during the flood (or high) tide; then it flows back out during the ebb (or low) tide. The tidal flow turns the blades of the turbine, which is connected to a generator that converts the kinetic energy of the turning blades into electricity. Tidal energy can also be captured in tidal streams, which are areas of high sea currents around headlands—bits of land that jut out into the water.

The main advantage of tidal power is that it is predictable; tides are both regular and well understood. Another advantage is that the technology to harness tides is already available and functional. Still, "Although the technology required to harness tidal energy is well established, tidal power is expensive, and there is only one major tidal generating station in operation." The one working tidal power station is located at the mouth of the La Rance River estuary in northern France, and it generates 240 megawatts of energy, about a quarter of the power one would expect from a major coal or nuclear power plant. "The La Rance generating station has been in operation since 1966 and has been a very reliable source of electricity for France."[2] The station was intended to be the first of many tidal power generating stations, but France chose instead to develop its nuclear power capacity.

There are currently some plans to build tidal power stations in Britain. In particular, the government has hoped to place a tidal barrage across the Severn estuary in Wales. The Severn has a very large difference between high and low tides, and as a result a barrier across it could generate as much energy as eight coal plants. This would significantly reduce Britain's greenhouse gas

The facility at the mouth of France's La Rance River uses massive turbines to extract power from the tides. Marcel Mochet/AFP/Getty Images.

Water and Ice

emissions. Environmentalists are divided on the project, however, because they fear that "the barrage would destroy vast areas of mudflats and marshes, which are vital feeding grounds for tens of thousands of wading birds, and [would] prevent migratory fish such as salmon and eels from ascending rivers to spawn."[3]

Wave Power

Another way to harness water energy is by using ocean waves. Unlike tidal energy technology, wave energy technology is still mostly in the development phase. Thus, "ocean-power systems now come in a staggering array of shapes, sizes and configurations," according to science reporter Sandi Doughton.[4] For example, one wave energy system design involves a chamber open to the waves with a hole containing a turbine in the top. As the water in the chamber rises, air is forced out through the hole, turning the turbine. Another method involves a long, hinged tube filled with fluid. The tube is placed in the ocean, and the waves bend it back and forth, pumping out the fluid, which can then power generators. Scientists and engineers are still researching and testing to determine which, if any, of these designs will work. For instance, a wave energy buoy designed by a company called Finavera "sank unexpectedly during a test run" off the Pacific Coast, according to *Seattle Times* staff reporter Michelle Ma.[5] After other hold ups and difficulties, Finavera decided to scrap its wave energy project altogether and focus instead on wind energy.

Wave energy may have some environmental impacts. There are concerns that generators may harm marine habitats, and that there will be "conflict with other sea space users, such as commercial shipping and recreational boating."[6] For example, scientists are concerned that a planned wave energy generating station off the coast of Scotland may be a danger to whales, dolphins, and seals, which may "be unable to detect the underwater turbines in time . . . and will be injured or killed by the rotating 16m-wide blades."[7]

Like tidal energy, however, wave energy does not generate greenhouse gases. And though some places do not have high enough waves to generate energy, others seem perfect for wave technology. In the Pacific Northwest, for example, "Waves big enough to generate power occur 80 percent to 90 percent of the time."[8] This is far more reliable than the typical wind-powered generating system. Thus, while wind power technology is much more developed at the moment, wave power may yet become an important source of energy in some regions.

Hydroelectric Power: Pros and Cons

Whereas wave and tidal power are still in development, a kind of water power known as hydroelectricity is already widespread. Hydroelectricity refers to power generated by hydroelectric dams. These dams are placed across rivers, creating huge lakes or reservoirs of water behind the dams. The water can then be released as needed, pouring over turbines that generate power.

Hydroelectricity already generates massive amounts of electricity. Hydroelectric dams accounted for about 20 percent of total electricity worldwide in 2005, according to *Renewables Global Status Report 2006 Update*.[9] In the United States, "about 7 percent of total power is produced by hydroelectric plants." While "most of the good spots to locate hydro plants have already been taken" in the U.S., as much as two-thirds of potential hydropower remains to be developed worldwide in Latin America, Africa, India, and China.[10]

Hydroelectric power is a useful form of energy for several reasons. First of all, it is renewable. Hydroelectricity is essentially powered by the sun, which evaporates water. That evaporated water then falls as rain, filling the rivers that supply the reservoirs behind the dam. As long as water evaporates and falls as rain, the water behind the dam will be renewed, which means that there will always be more water to flow downstream and turn the turbines. In addition, dams "have a low operating cost once installed and can be highly automated."[11]

Water and Ice

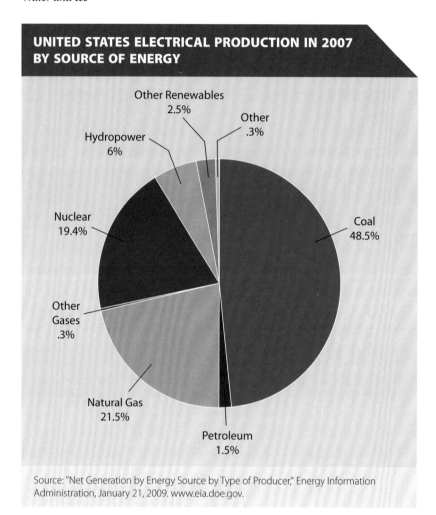

UNITED STATES ELECTRICAL PRODUCTION IN 2007 BY SOURCE OF ENERGY

- Other Renewables 2.5%
- Other .3%
- Hydropower 6%
- Nuclear 19.4%
- Coal 48.5%
- Other Gases .3%
- Natural Gas 21.5%
- Petroleum 1.5%

Source: "Net Generation by Energy Source by Type of Producer," Energy Information Administration, January 21, 2009. www.eia.doe.gov.

Another advantage of hydropower is that hydroelectric dams can vary the release of electricity easily, simply by reducing water flow during slow times and increasing it at peak hours. "By comparison, thermal power plants take much longer to start up when they are 'cold,'" according to Catherine Gautier in her book *Oil, Water, and Climate*.[12] Furthermore, damming rivers has many other advantages besides generating electricity. Dams are an important source of water for irrigation and for drinking water. Indeed, according to Gautier, "about half of the world's largest dams were built solely or primarily for irrigation."[13]

There are numerous disadvantages to the use of hydroelectricity as well. Dams can have large negative ecological effects. For example, dams often interfere with the migration patterns of fish. Thus, on the Mekong River in Southeast Asia, where "many commercial species [of fish] have highly developed migratory patterns . . . dams act as a barrier to fish migrating upstream; and returning fish migrating downstream . . . generally must go through turbines. . . . As a result, spawning is greatly reduced and replenishment of fish stocks is diminished," according to Chris Barlow, the manager of the Mekong River Commission's fisheries program.[14]

Falling fish stocks may cut into human food supplies. But dam construction can affect people directly as well. When a river is dammed, it creates a large lake. This lake may flood human settlements, forcing massive migration. For example, one proposed dam in northwestern Burma on the Chindwin River would generate more electricity than all of Burma's current generating capacity combined—but it would also flood 17,000 acres (6.880 hectares) and force "30,000 inhabitants to move."[15]

Dams also prevent rivers from flooding, which can have serious consequences on wetlands and on groundwater tables around the river. Vegetation in a river's floodplain may dry out, and species that depend on regular floodwaters may be threatened. Furthermore, according to Catherine Gautier, dams cause "river flows to fluctuate unnaturally" as water is released and contained in accordance with electrical needs. These fluctuations can wreak havoc with the stable water temperatures and constant oxygen levels that fish species need to survive.[16]

There has also been concern that dam building may cause earthquakes. Some scientists have suggested that the massive Chinese earthquake of May 2008, which killed 80,000 people, may have been triggered in part by water collected behind the massive Zipingpu Dam. "The dam was built 500 meters from the earthquake's fault line. A research paper by a group of Chinese scientists concluded that the weight of collected water clearly affected seismic activity," according to a 2009 *Wall Street Journal* article.[17]

Hydroelectric Power and Greenhouse Gases

In the debate about hydroelectric power, one of the most contentious issues involves greenhouse gases. Like tidal and wave power, hydroelectric power does not burn fuel and so does not generate carbon dioxide emissions directly. Some commentators have therefore argued that hydroelectric power would reduce greenhouse gas emissions and moderate climate change. For example, Kim Murphy in a 2009 *Los Angeles Times* article noted that "The ability of the nation's aging hydroelectric dams to produce energy free of the curse of greenhouse gas emissions and Middle Eastern politics has suddenly made them financially attractive—thanks to the economics of climate change."[18]

There is some question, however, about whether hydroelectric dams do in fact reduce greenhouse emissions. When a dam is built, the reservoir created behind it floods a large land area. The result is that a large number of land plants end up underwater. Because the water levels of the lake continually rise and fall, new plant matter on the banks is continually growing and then being submerged. This means that the lake is continually being supplied with dead plant matter. This dead plant matter is broken down by bacteria underwater, where there is no oxygen, in a process that produces a gas called methane.

The methane dissolved in the water is held in solution by the high pressure of all the water above it. Thus, in normal circumstances, the methane would enter the atmosphere only slowly, through the surface of the lake. When the dam gates are open, however, and the water rushes through, that pressure is suddenly relieved. The effect is like opening a bottle of Coca-Cola, with gas

Water and Energy

The Vakhsh River and Climate Change

Tajikistan is a mountainous, land-locked country in central Asia. It borders Afghanistan, Uzbekistan, and Kyrgyzstan, and lies close to Pakistan. The largest river in Tajikistan is the Vakhsh. The Vakhsh originates in the southwestern, glacier-covered mountains of Kyrgyzstan, and it crosses the entire length of Tajikistan before joining with the Panj River to form the Amu Darya at the border of Tajikistan and Afghanistan.

The Vakhsh is Tajikistan's major source of electricity. Five dams are currently in place on the Vakhsh, including the Nurek Dam, which, as of 2009, is the tallest dam on earth. The Nurek includes nine hydroelectric turbines, which generate 3 gigawatts of power. Altogether, the five hydroelectric dams on the Vakhsh generate 90 percent of Tajikistan's electricity. This output has not been enough to keep up with Tajikistan's electrical needs, however, and the nation has been forced to ration electricity at times. To increase capacity, four more dams are planned along the river. One of them, the Rogun Dam, will be taller than the Nurek and is expected to generate 3.6 gigawatts of power. In addition to the electricity it generates, the Vakhsh is vital as a source of irrigation for agriculture.

Because it is so centrally important to Tajikistan, anything that threatens the Vakhsh is a threat to the entire country. One danger is from earthquakes; landslides have occasionally blocked portions of the river and damaged dam equipment. Another potential long-term threat is global warming. The Vakhsh originates in mountain glaciers. If temperatures warm enough to reduce the size of those glaciers significantly, the flow of Tajikistan's major electricity source might be changed irreversibly.

bubbles rushing to the surface . . . except that "The difference in pressure between a closed and open bottle of Coca-Cola is minor compared to the pressure at depth in a hydroelectric reservoir," according to researcher Philip M. Fearnside.[19] The methane that bubbles out of the water enters the atmosphere—where it contributes to climate change.

In fact, methane "is estimated to be more than 20 times as powerful," as carbon dioxide in contributing to climate change.[20] Thus, hydroelectric dams essentially function as gigantic engines for taking the relatively low-impact carbon dioxide stored in plant matter, changing it into high-impact methane, and releasing it quickly into the atmosphere. The problem is exacerbated in "tropical regions where temperatures are high and decomposition is more rapid," according to Gautier.[21] Some environmentalists have even argued that the greenhouse emissions from a hydroelectric dam do more damage than ones from a comparable coal plant.

Whether this is true or not is difficult to determine. Gautier notes that, right after they begin to function, dam emissions are especially high, "as a result of the decomposition of flooded vegetation and soils, as well as construction-related deforestation." After that, the amount of methane emitted varies from dam to dam. Dams with deep, narrow reservoirs produce fewer emissions relative to their power output than coal plants do. Reservoirs that are broader and shallower, however, may produce more emissions per watt than coal plants do.[22]

Some researchers are working on an innovative means of increasing the efficiency of the dams. Methane does not have to be a waste product; it could itself be a source of energy. If the methane released by the dams could be captured and burned, it would release a huge amount of energy. In fact, "Some hydroelectric plants in the Amazon hold an added energy capacity of 27 to 53 percent, taking into account the methane bubbles released from the water as it passes through the turbines and spillways."[23] Brazil's second largest hydroelectric dam at Tucurui could yield one million metric tonnes (907,000 tons) of methane. When burned, that amount of methane would release enough energy to make the construction of several other hydroelectric dams unnecessary. Scientists are still working on perfecting a technique to capture the methane. If they succeed and implement the technology worldwide, it "could prevent emissions equivalent to more than the total annual burning of fossil fuels in the UK," according to reporter Tim Hirsch.[24] Thus, in the future, hydroelectric dams, tidal power, and wave power may all make important contributions to the moderation of climate change.

Notes

1. National Resources Defense Council, "Global Warming Basics," October 18, 2005. www.nrdc.org.
2. Renewable Energy Office for Cornwall, "Tidal Power," n.d. www.reoc.info.
3. Michael McCarthy, "The Great Divide: Green Dilemma over Plans for Severn Barrage," *Independent*, January 27, 2009. www.independent.co.uk.
4. Sandi Doughton, "Tapping Tidal Energy: The Wave of the Future," *Seattle Times*, October 7, 2007. http://seattletimes.nwsource.com.
5. Michelle Ma, "Wave-Energy Project Halted," *Seattle Times*, February 11, 2009. http://seattletimes.nwsource.com.
6. *OCS Alternative Energy and Alternate Use Programmatic EIS*, "Ocean Wave Energy," n.d. http://ocsenergy.anl.gov.
7. Mark Macaskill, "Sea Mammals 'at Risk from Wave Turbines,'" *Sunday Times*, June 14, 2009. www.timesonline.co.uk.
8. Sandi Doughton, *Seattle Times*.
9. REN21, *Renewables Global Status Report 2006 Update*, 2006, p. 20. www.ren21.net.
10. *Georgia Water Science*, "Water Use: Hydroelectric Power," May 13, 2009. http://ga.water.usgs.gov.
11. Environmental Literacy Council, "Hydroelectric Power," n.d. www.enviroliteracy.org.
12. Catherine Gautier, *Oil, Water, and Climate: An Introduction*. New York: Cambridge University Press, 2008, p. 214.
13. Gautier, *Oil, Water, and Climate*, p.214.
14. Chris Barlow, "Dams, Fish, and Fisheries in the Mekong River Basin," *Mekong River Commission*, n.d. www.mrcmekong.org.
15. Yeni, "Hydro Dam May Force 30,000 to Move," *Irrawaddy*, August 30, 2006. www.irrawaddy.org.
16. Gautier, *Oil, Water, and Climate*, p. 214.
17. Gautam Naik and Shai Oster, "Scientists Link China's Dam to Earthquake, Renewing Debate," *Wall Street Journal*, February 6, 2009. http://online.wsj.com.
18. Kim Murphy, "Boom in Hydropower Pits Fish Against Climate," *Los Angeles Times*, July 27, 2009. www.latimes.com.
19. Philip M. Fearnside, "Greenhouse Gas Emissions from Hydroelectric Dams: Controversies Provide a Springboard for Rethinking a Supposedly 'Clean' Energy Source," *Climatic Change* 66(1-2), 2004. http://philip.inpa.gov.br.
20. Tim Hirsch, "Project Aims to Extract Dam Methane," *BBC News Online*, May 10, 2007. http://news.bbc.co.uk.

Water and Ice

21. Gautier, *Oil, Water, and Climate*, p. 217.
22. Gautier, *Oil, Water, and Climate*, p. 218.
23. Mario Osava, "Brazil: Tapping Hydroelectric Dams for Methane Gas," *IPS News*, June 6, 2007. www.ipsnews.net.
24. Hirsch, "Project Aims to Extract Dam Methane."

Chapter 7

Conclusion

Water Affects Climate

Water affects the climate in many ways. One of the most important of them is through the hydrologic cycle. The hydrologic cycle is the circulation of water molecules from the earth's surface to the atmosphere and back again. The hydrologic cycle is powered by the sun, which heats the earth, turning water on the surface to water vapor. The water vapor then rises into the atmosphere, where it condenses and falls back to the earth as rain. Local and seasonal variations in the hydrologic cycle can have a major effect on climate.

The world's oceans also have a major effect on climate. Water has a higher specific heat than land, which means that it is more resistant to changes in temperature than is land. Thus, the fact that the majority of the earth is covered in water helps to keep temperatures stable worldwide. Moreover, the high specific heat of water affects local climates. Regions of land that are close to water tend to have warmer winters and cooler summers. Regions of land that are farther from water tend to have cold winters and hot summers.

Because water is so important to climate, the rise in temperature caused by global warming will affect the earth's oceans and water supply. At the same time, water on the earth affects the pace of climate change.

Water and Ice

The ocean plays a particularly important role in moderating climate change. First, the ocean has a higher specific heat than land, which means that its temperature rises more slowly than that of land. Effectively, this means that the ocean slows global warming.

Second, the ocean absorbs carbon from the atmosphere. Thus, much of the carbon dioxide that humans release into the air from burning fossil fuels is absorbed and contained by ocean waters. Again, this means that the ocean slows global warming.

The effectiveness of the ocean in slowing global warming may be reduced over time, however. As water temperatures rise, the ocean becomes less able to absorb carbon dioxide. As global warming accelerates, the ocean will become a less effective brake, which will accelerate global warming further.

This kind of effect—wherein a given phenomenon creates conditions that increase or exacerbate the phenomenon—is called a feedback loop. Scientists are concerned about a number of feedback loops involving water and ice.

One of the most important of these phenomena is a water vapor feedback loop. Water vapor is itself a greenhouse gas; that is, more water vapor in the atmosphere tends to trap heat on the earth and raise temperatures. As temperatures go up, more water on the earth evaporates, becoming water vapor and going into the atmosphere—where it raises temperatures. Scientists are not sure, however, how important this effect will be in the long term. In part, they are uncertain because more water vapor in the atmosphere might also mean more cloud cover, which would reflect sunlight and cool the earth.

Another potentially important feedback loop involves ice albedo. Albedo is the ability of a substance or body to reflect sunlight. Ice has a very high albedo; ocean water has a much lower albedo. Scientists worry that, as temperatures rise worldwide, ice may melt. With less ice, more ocean water will be exposed. That means that the overall albedo of the ocean will fall, and the ocean will absorb more heat faster—possibly raising the temperature of the water further and melting more ice.

Sea Levels, Rain, and Drought

The rising temperature of ocean water has a number of important consequences. Specifically, as water heats, it expands. This is referred to as thermal expansion. Many scientists believe, therefore, that as ocean temperature rises, the water will take up more space, and sea levels will rise.

Global warming may contribute to a rise in sea levels in other ways as well. As world temperatures rise, large, land-based ice flows called glaciers may begin to melt. This outcome could be especially significant if the massive ice sheets on Antarctica and Greenland were to begin to melt.

Predicting sea-level rise is very difficult, and scientists disagree about how much exactly sea levels can be expected to rise. Some argue that sea-level rise may not be significant. Others worry that a melting of the Greenland ice sheet could mean a rise of 10 to 20 feet (3–6m) in sea level. The general consensus among scientists, however, is that sea-level rise will be between these two extremes. The Intergovernmental Panel on Climate Change (IPCC) predicts that sea-level rise will be between 7 and 23 inches (18–59cm) over the next century.

A significant rise in sea level would affect human beings in numerous ways. Coastal areas would lose land and be subject to increased flooding from storms. Regions that might be affected include the east coast of the United States, California, Nigeria, the Netherlands, and other low-lying areas worldwide. In general, the areas that would be hardest hit are poor regions such as Bangladesh, which do not have the resources to prepare or react to major changes.

Rising sea levels may have similar effects on islands, which could lose land or, in some cases, be submerged altogether. This possibility is a particular concern on some small Pacific islands, though the exact extent of the danger is difficult to assess.

Water and Ice

After rising sea level, perhaps the biggest concern about how global warming may affect world water involves rain. Warm air holds more water vapor than cold air, so scientists expect that rainfall worldwide should increase by 3 percent to 5 percent, though such predictions are speculative at best.

Increased rainfall would have a number of benefits. In the first place, more rain might mean more cloud cover, which could slow global warming. Moreover, rain is good for crops and agriculture. More water in the atmosphere might also mean more extreme weather events like hurricanes, however.

In addition, though the world overall should be wetter, some areas may in fact become drier as a result of global warming. For example, the increase in evaporation rates may dry out soils, turning rain forests to savannahs in Latin America.

Warmer temperatures and higher evaporation rates could also contribute to droughts. Scientists are particularly concerned that global warming may be shrinking rivers and lakes around the world. Global warming may also result in more rain and less snowfall, which could have a dangerous effect on areas such as the American Southwest, which rely on melting snow for water during the dry summer months.

Energy from Water

Although scientists are concerned about the impact of global warming on world water supplies, they are also hopeful that energy from water may help to moderate climate change. There are several ways to harness the power of water to produce clean energy without the burning of fossil fuels. Among them are tidal power and wave power.

Tidal power is expensive, and wave power technology is not fully developed. In contrast, hydroelectric power is well established. Hydroelectric power involves building dams across rivers. Once built, hydroelectric dams are cheap to run, and their energy is renewable, as long as rain refills the rivers.

Hydroelectric dams raise many environmental concerns, however. Building dams can damage ecosystems. Moreover, though they do not release carbon dioxide, hydroelectric dams that flood forests may create large amounts of methane, which is itself a powerful greenhouse gas. For these reasons, the use of hydroelectric dams remains controversial.

The Future of Water and Climate

Although disagreement exists about the extent to which hydroelectric power could help to mitigate global warming, there is no doubt that water and ice will play a central role in climate change. Many of the most serious fears about climate change—rising sea levels, increased storm activity, droughts—involve the effects of global warming on water and ice. Similarly, the many hopes for mitigating global warming—through hydroelectric power, or through the cultivation of wetlands—involve the planet's water. In short, climate and water on the earth are so closely connected that it is impossible to alter one without affecting the other.

Glossary

albedo A measure of the amount of sunlight an object reflects.

atoll A coral island encircling a lagoon.

carbon dioxide (CO_2) A greenhouse gas chemically composed of carbon and oxygen. Carbon dioxide is produced by the burning of fossil fuels such as oil and coal. It is also exhaled by human beings and inhaled by plants. Burning plants also releases carbon dioxide.

continental climates Climates that are relatively cold in winter and hot in summer; usually found at the interiors of continents.

El Niño A periodic oceanic warming, which can have major effects on weather.

feedback loop A process in which one condition creates other conditions that reinforce the first.

glacier A mass of ice that year-round is located on and moves over land.

global warming The increase in the average temperature of the earth's surface and oceans. Global warming has been occurring since the mid-twentieth century, and it is expected to continue because of the greenhouse effect.

greenhouse effect The heating of the surface of the earth due to the presence of gases that trap energy from the sun.

greenhouse gases Substances that contribute to the greenhouse effect and global warming. Carbon dioxide, methane, and water vapor are all greenhouse gases.

groundwater Water located underneath the ground, either in porous soil or in spaces in rock.

Hurricane Katrina The 2005 storm that devastated the city of New Orleans.

hydroelectric power Electrical power generated by dams.

hydrologic cycle The circulation of water from the earth's surface as a solid or liquid to the air as water vapor and back to the earth as precipitation.

hydropower Power derived from the motion of water.

ice sheet A mass of glacial ice more than 20,000 square miles (50,000 sq. km) in area.

ice shelf A floating platform of ice that forms at the junction of a glacier and a coastline.

Intergovernmental Panel on Climate Change (IPCC) A scientific body established by the United Nations to evaluate the risk of climate change caused by human activity.

jet stream Fast flowing, narrow air currents in the atmosphere.

mangroves Trees and shrubs that grow in salt-water habitats.

maritime climates Climates that are relatively warm in winter and mild in summer; usually located near the ocean.

methane A greenhouse gas about 20 times more potent than carbon dioxide. Methane gas is produced when organic matter decays in the absence of oxygen. It also exists below the earth's surface and is produced in the intestines of livestock, such as cows, which release it into the atmosphere through flatulence.

monsoons Seasonal winds, which often have a profound effect on seasonal precipitation.

moulin A narrow, tubular chute, hole, or crevasse through which water enters a glacier from its surface.

Water and Ice

North Atlantic Current An important ocean current in the Atlantic that moves warm surface waters from the equator north toward the Arctic, then sinks down and pulls cold water back toward the equator.

snowpack Naturally formed compressed snow. It is especially common in mountainous regions, and it often melts in springtime.

specific heat A measure of the amount of energy needed to increase the temperature of a substance by a set amount.

sublimation The process whereby snow or ice turns directly into water vapor.

thermal expansion The increase in volume in matter caused by an increase in temperature. Thermal expansion of ocean water as a result of global warming is believed to be one of the major short-term causes of sea-level rise.

tidal power The conversion of tidal energy into useful forms of power.

transpiration The release of water from plants.

wave power The conversion of wave energy into useful forms of power.

wetlands An area of land whose soil is saturated with moisture, either permanently or seasonally.

For Further Research

Books

David Archer, *The Long Thaw: How Humans Are Changing the Next 100,000 Years of Earth's Climate*. Princeton, NJ: Princeton University Press, 2009.
> Argues that human-caused global warming will have major short-term and especially long-term effects.

Anthony D. Barnosky, *Heatstroke: Nature in an Age of Global Warming*. Washington, DC: Island Press, 2009.
> Argues that global warming is having a devastating impact on animal and plant ecosystems.

Grant Bigg, *The Oceans and Climate,* 2nd ed. New York: Cambridge University Press, 2003.
> Discusses the relationship between ocean and climate from a scientific perspective.

Gary Braasch, *Earth Under Fire: How Global Warming Is Changing the World*. Berkeley, CA: University of California Press, 2007.
> Argues through words and numerous full-color photographs that global warming is having a major impact on the earth.

Christopher Essex and Ross McKitrick, *Taken by Storm: The Troubled Science, Policy, and Politics of Global Warming*, revised edition. Toronto, Ontario: Key Porter Books, 2007.
> Argues that evidence for global warming as a serious problem is weak.

Catherine Gautier, *Oil, Water, and Climate: An Introduction*. New York: Cambridge University Press, 2008.
> Discusses the scientific, economic, and policy connections between energy, water, and climate issues.

Al Gore, *An Inconvenient Truth: The Planetary Emergency of Global Warming and What We Can Do About It*. New York: Rodale Books, 2006.
> Argues that the evidence for global warming is clear and that government action to address the problem is imperative.

Bill McGuire, *Seven Years to Save the Planet: The Questions . . . and Answers*. London, UK; Weidenfeld & Nicholson, 2008.
> Argues that immediate and sweeping action is needed to prevent devastating climate change.

Patrick J. Michaels, *Meltdown: The Predictable Distortion of Global Warming by Scientists, Politicians, and the Media*. Washington, DC: Cato Institute, 2004.
> Argues that the evidence for global warming as a serious problem is weak, and suggests that the scientific establishment is biased in favor of climate change.

James Lawrence Powell, *Dead Pool: Lake Powell, Global Warming, and the Future of Water in the West*. Berkeley, CA: University of California Press, 2008.
> Argues that water usage and policy in the American Southwest is unsustainable, especially in light of global warming.

Andrew Weaver, *Keeping Our Cool: Canada in a Warming World*. Toronto, Ontario: Penguin, 2008.
> Debunks some global warming myths while arguing that climate change is a serious problem that Canada must address forcefully.

Periodicals

David Adam, "Global Warming May Trigger Carbon 'Time Bomb,' Scientist Warns," *Guardian*, March 10, 2009.

Jessica Aldred, "Global Warming 'Changing Balance' of Marine Life in Polar Seas," *Guardian*, February 15, 2009.

Allen Best, "Thirsting for Water," *Forest Magazine*, Summer 2007.

Sandi Doughton, "Tapping Tidal Energy: The Wave of the Future," *Seattle Times*, October 7, 2007.

Greg Easterbrook, "Global Warming: Who Loses—and Who Wins," *Atlantic*, April 2007.

Economist, "Thaw Point," July 30, 2009.

Philip M. Fearnside, "Greenhouse Gas Emissions from Hydroelectric Dams: Controversies Provide a Springboard for

Rethinking a Supposedly 'Clean' Energy Source," *Climatic Change* 66(1-2), 2004.

Roger Highfield, "*An Inconvenient Truth* Exaggerated Sea Level Rise," *Daily Telegraph*, September 4, 2008.

Michelle Ma, "Wave-Energy Project Halted," *Seattle Times*, February 11, 2009.

Andrew Marshall, "Treading Water," *Time*, September 28, 2009.

Michael McCarthy, "The Great Divide: Green Dilemma over Plans for Severn Barrage," *Independent*, January 27, 2009.

Curtis A. Moore, "Awash in a Rising Sea: How Global Warming Is Overwhelming the Islands of the Tropical Pacific," *International Wildlife*, January-February 2002.

Kim Murphy, "Boom in Hydropower Pits Fish Against Climate," *Los Angeles Times*, July 27, 2009.

Gautam Naik and Shai Oster, "Scientists Link China's Dam to Earthquake, Renewing Debate," *Wall Street Journal*, February 6, 2009.

Alan Prendergast, "The Skeptic," *Denver Westword News*, June 29, 2006.

Stefan Rahmstorf, "We Must Shake Off This Inertia to Keep Sea Level Rises to a Minimum," *Guardian*, March 3, 2009.

Andrew C. Revkin, "Global Warming Could Forestall Ice Age," *New York Times*, September 3, 2009.

Richard Seager, "The Source of Europe's Mild Climate," *American Scientist*, August 2006.

Marco Sibaja, "Climate Change Threatens Brazil's Coffee Crop," *USA Today*, February 19, 2009.

Jim Tankersley, "California Farms, Vineyards in Peril from Warming, U.S. Energy Secretary Warns," February 4, 2009.

Water and Ice

Gerald Traufetter, "Global Warming a Boon for Greenland's Farmers," *Der Spiegel*, August 30, 2006.

Bryan Walsh, "Is Global Warming Worsening Hurricanes?" *Time*, September 8, 2008.

Internet Sources

CBC News Online, "Kilimanjaro Glaciers Could Go Within Decades," November 2, 2009. www.cbc.ca.

Ed Corcoran, "Addressing Global Warming: Water," Global Security.org, July 29, 2009. http://sitrep.globalsecurity.org.

Lisa Gardiner, "Ice-Albedo Feedback: How Melting Ice Causes More Ice to Melt," Windows to the Universe, July 18, 2007. www.windows.ucar.edu.

Georgia Water Science, "Water Use: Hydroelectric Power," May 13, 2009. http://ga.water.usgs.gov.

Steve Graham, Claire Parkinson, and Mous Chahine, "The Water Cycle," Earth Observatory, n.d. http://earthobservatory.nasa.gov.

Natural Resources Defense Council, "The Consequences of Global Warming," December 10, 2008. www.nrdc.org.

Renewable Energy Office for Cornwall, "Tidal Power," n.d. www.reoc.info.

Randy Russell, "Global Warming, Clouds, and Albedo: Feedback Loops," Windows to the Universe, May 17, 2007. www.windows.ucar.edu.

UNEP Grid-Arendal, "Climate Change 2001: Working Group 1: The Scientific Basis," 2003. www.grida.no.

Union of Concerned Scientists, "Early Warning Signs of Global Warming: Droughts and Fires," November 10, 2003. www.ucsusa.org.

University Corporation for Atmospheric Research, "Water Levels Dropping in Some Major Rivers as Global Climate Changes," April 21, 2009. www.ucar.edu.

U.S. EPA, "Coastal Zones and Sea Level Rise," August 3, 2009. www.epa.gov.

Web Sites

Cato Institute (www.cato.org) The Cato Institute is a libertarian think-tank; it is concerned that climate change regulation will create undue economic burdens. Book chapters, policy papers, studies, and audio files are available.

Climate Change—U.S. EPA (www.epa.gov/climatechange) The EPA (Environmental Protection Agency) is the U.S. government body charged with environmental regulation. Basic science information, reports on regulatory initiatives, and news releases are available.

IPCC—Intergovernmental Panel on Climate Change (www.ipcc.ch) The IPCC is the main international body devoted to research into human effects on climate change. Reports, working papers, news, data, and press releases are available.

Union of Concerned Scientists (UCS) (www.ucsusa.org) The UCS is a nonprofit science advocacy group concerned mostly with environmental issues, such as confronting global warming. Policy and analysis articles are available.

Index

A

Abeta, Teunaia, 54, 55
Adam, David, 20
Adaptation, 7, 84, 85
Aerosol, 3, 7
Afghanistan, 95
Africa, 54, 67, 70, 78, 82, 91
Agriculture
   in Bangladesh, 54
   farming practices, 4, 7
   groundwater for, 81
   human adaptation in, 84–85
   rainfall and, 16, 62
   temperature change and, 71, 82
   See also Irrigation
Albedo, 24, 25, 26, 32, 37, 104
Alternative energy, 8
Amazon rain forest, 71, 96
American Scientist (magazine), 41
Amu Darya River (Asia), 95
An Inconvenient Truth (documentary film), 43, 44, *45*, 63
Andes mountains (South American), 81
Antarctic ice sheet, 35, 36*t*, 38
Antarctica, 32, 34, 35, 37, 101
Aquaman (superhero), 42
Aquifers, 49, 81
Arab militias, 69
Archer, David, 17, 19, 22, 23, 35–37, 59, 67
Arctic, 6, 26, 41, 75
Asia, 71, 95
Atlantic Coast, 47*t*
Atlantic hurricanes, 6, 66
Atlantic Ocean, 43
Atmosphere, 4, 20, 23, 99
Atmospheric convection, 63
Atolls, 55, 104
Aumann, Hartmut, 67
Austin, Jay, 76
Australia, 28, 44, 55
Awudi, George, 57

B

Bangladesh, 53–54, 56, 59, 60, 101
Barlow, Chris, 93
Barnett, Tim, 79, 81
Barnosky, Anthony D., 71
Battisti, David, 82
Bello, Muhammadu, 77
Benford, Gregory, 21
Bicarbonate, 22
Bigg, Grant R., 24
Biofuels, 21
Black carbon, 28
   See also Carbon
Boca Raton (FL), 42
Bolivia, *15*
Braasch, Gary, 70
Braithwaite, Roger, 33
Brazil, 84–85, 96
Brest (France), 31
Britain, 60, 89
Brown, Lester R., 37
Brushfires, 71
Bureau of Reclamation, *83*
Burma, 93
Burps, 43
"Business as usual" scenario, 6
Butzengeiger, Sonja, 54

C

Caldeira, Ken, 66
California, 50*t*, 52–53, 56, *83*, *101*
Cameroon, 77
Canadian Wildlife Service, 33
Carbon
   black, 28
   in oceans, 20–22, 43
   uptake, 22, 100
   in wetlands, 51
Carbon dioxide (CO2)
   definition, 104
   forest fires and, 26–27
   fossil fuels causing, 87
   methane and, 96
   as nutrient, 62

Index

oceans and, 20–23, 100
   trees and, 26
Carbonate ions, 22
Cargo ships, 76
Cascade Mountains, 79
Cemeteries, 54
Central Valley, California, 75
Chad, 77
Chicago (IL), 19
China, 84, 91
Chindwin River (Burma), 93
Cleveland (OH), 13
Climate change
   Darfur slaughter and, 69
   intervention and, 7, 8
   models and, 5
   precipitation patterns and, 74–75
   water and, 99
   See also Global warming; Warming
Climate Change 2007 (report), 81
Climate disaster, 42
Climate models, 4, 5, 28
Climates, 67–72
Cloud albedo feedback loop, 24
Clouds, 12, 23, 24, 67
CO_2. *See* Carbon dioxide (CO_2)
Coal, 4, 87, 96
Coal-burning power plants, 87
Coastal regions
   American, 7, 43–44, 50–52
   Australian, 44
   global, 53–54, 101
Coastal Sensitivity to Sea-Level Rise (U.S. government report), 56
Coca-Cola, 94–95
Coffee, 84–85
Cohen, Andrew, 77, 78
Colorado River, 74, 79
Columbia River, 74
Communities, 56–57
Components, 3
Computer models, 35–36, 63, 79
   See also Models
Condensation, 20
Congo, 60
Conservation, 8
Continental climates, 14, 67, 104

Continental drift, 3
Cook, John, 23
Corn, 21
Crop residue, 21
Crops, 54, 82, 84, 85
Currents, 40, 41, 52, 75
Curry, Ruth, 40

D

Dai, Aiguo, 75
Daily Mail (newspaper), 33, 95
Dallas (TX), 61
Dams
   ecosystems and, 93–94, 103
   need for, 81
   in Netherlands, 53
   pros/cons of, 91–94, 96, 102
   river flow reduction from, 74
   in Tajikistan, 95
   See also Dikes
Darfur (Africa), 69
The Day After Tomorrow (film), 42
Dead Pool (Powell), 79
Dell, John, 42
Democratic Republic of Congo, 60
Deserts, 27
Dettinger, Michael, 81
Development, 51
Diaz, Henry, 35
Diffenbaugh, Noah, 16
Dikes, 49, 52, 53, 56
Disease, 7
Domingues, Catia, 31
Doughton, Sandi, 90
Drainage, 49, 51, 56
Drinking water, 14, 16, 49, 60, 92
Drought, 69, 70, 71, 82, 83, 102
Dust, 27
Dust Bowl, 70

E

Earth Observatory (Web site), 11
Earth Policy Institute, 37
Earth Under Fire (Braasch), 70
Earthquakes, 93, 94, 95
El Niño, 4, 15–17, 66, 104
Electrical production, 92*t*
Electricity, 88, 91, 92, 93, 95

113

Emissions, 5
Emissions reduction, 8, 21
End-of-the-world scenarios, 42
Energy sources, *92*
England, 60–61, 62
English Channel ferries, 42
Environment, 7, 90
Environment Canada, 60
Equator, 20, 66
Erosion, 3, 55, 56
Essex, Christopher, 44, 56, 63, 66, 84
Estuaries, 88, 89
Europe, 6, 41, 42, 82
Evaporation
    hurricanes and, 63
    in hydrologic cycle, 10, 11
    ice and, 76
    from oceans, 20, 44
    in rain forest, 71
Extinction, 33

F

"Farewell Tuvalu" (article), 55
Farmers, 69
    See also Agriculture
Farming. *See* Agriculture
Fearnside, Philip M., 95
Feedback effects, 28, 37
Feedback loops
    defined, 100, 104
    ice albedo, 24–26
    negative, 62
    water/climate and, 22–23
    water vapor, 23–24, 28
    for wetlands, 51
Ferries, 42
Finavera (energy company), 90
Fish/fishing, 77, 78, 93
Flooding
    in Bolivia, 15
    California population and, 50*t*, 52–53
    damage from, 49, 55
    dams and, 93, 103
    from glacial melting, 35
    global warming and, 67
    from monsoons, 16
    protection from, 49

wetlands and, 51
Florida, 49, 52
Food supplies, 7, 77, 78
Forest fires, 26
Fossil fuel emissions, 5, 6
Fossil fuels, 4, 21, 87, 97, 100
France, 82, 89
Freshwater, 41, 75, 78
Friends of the Earth, 57

G

Ganges River (India), 74
Garber, Kent, 85
Gardiner, Lisa, 26
Gautier, Catherine, 63, 66, 93, 96
    Oil, Water, and Climate, 63, 78, 92
Generators, 90
Genocide, 69
Geoengineering, 7
Geologic uplift, 56
Georgia Tech (GA), 17
German Aerospace Centre, 77
Germanwatch (policy organization), 54
Ghana, 60
Glaciers
    definition, 104
    melting of, 32–37, 41, 81, 101
    sea-level rise and, 32–34, 36*t*
Global cooling, 24
Global warming
    carbon absorption and, 20–22
    climate/precipitation patterns of, 24, 61
    definition, 104
    in Europe, 42
    feedback effects in, 26, 27, 28, 51
    monsoon season and, 16
    mountain glacial melt and, 34–35
    oceans and, 100
    in popular culture, 42
    rainfall increase and, 62, 102
    sea level rise and, 43–48, 100, 101
    snowpack and, 78–79
    in Sudan, 69
    See also Climate change; Warming
Gloucestershire (England), 60
Gore, Al, 44, *45*

An Inconvenient Truth, 43, 44, 45, 63
Government policies, 8
Great Britain, 60, 89
Great Lakes, 11, 76
Great Plains, 67
Greece (ancient), 87
Greenhouse effect, 27, 104
Greenhouse gases
    definition, 104
    hydroelectric power and, 94
    reducing, 7–8, 57
    warming effect of, 3, 4, 5, 7
    in wetlands, 51
Greenland
    albedo effect and, 37
    glaciers and, 32
    ice sheet, 6, 34–36, 43, 101
    sea level rise and, 39, 48
Groundwater, 81–82, 93, 104
Growing seasons, 84, 85
Gulf Coast, 7
Gulf Stream, 4
The Guardian (newspaper), 20

H

Hansen, James, 5, 27, 62
Hawaii, 55
Heat, 19, 20
Heat absorption, 16, 20, 31–32
Heat waves, 6
Heatstroke (Barnosky), 71
Henderson (NV), 61
Himalayan ice melt, 81
Hirsch, Tim, 97
Horstmann, Britta, 54
Human factors, 4, 5
Human health, 14, 15
Humidity, 10, 13, 23, 24, 66
Hurricane Katrina, 63, *64–65,* 66, 105
Hurricanes
    Atlantic, 6, 66
    El Niño and, 15–16
    global warming and, 66–67
    intensity changes in, 17, 63, *64–65*
    rainfall increases and, 102
Hydroelectric power
    definition, 105
    greenhouse gases and, 94–97
    monsoons and, 16
    pros/cons of, 91–94, 102–103
Hydrologic cycle, 10–13, 14, 15, 99, 105
Hydropower, 87–88, 105

I

Ice
    albedo, 24–26, 28, 100, 104
    cover, 76
    in lakes, 76
    melt, 32, 37, 41, 100, 101
    in sea/oceans, 32, 37, 46
    shelves, 37, *38*
Ice age, 40, 41, 43
Ice sheets, 19, 30–39, 101, 105
Ice-shelf, 37, 105
Icebergs, 32, 37
India, 16, 59, 84, 91
Indian Ocean, 70
The Independent (newspaper), 60
Industrial revolution, 4
Inertia, of climate system, 6
Infectious disease, 7
Insurance, 56
Intergovernmental Panel on Climate Change (IPCC), 30, 44, 45, 82, 101, 105
Irrigation
    canals, 82, *83*
    changes in, 82–84
    dams providing, 92–93, 95
    Lake Chad and, 77
    river flow reduction from, 74
    snowmelt and, 79
    in Western nations, 70
Islands, 54, 101
Italy, 82

J

Jet streams, 66, 105
JLA: American Dreams (Morrison), 42
Journal of Climate, 74

K

Kaser, Georg, 34
Katrina. *See* Hurricane Katrina

Keeping Our Cool (Weaver), 42, 61
Kilimanjaro, 34, 35
Kinshasa (Democratic Republic of Congo), 60
Kluger, Jeffrey, 63
Krygyzstan, 95

L

La Rance generating station, France, 89
Lagos (Nigeria), 54, 56
Lake Chad (West Africa), 77
Lake-effect snow, 13
Lake Huron, 76
Lake Michigan, 76
Lake Superior, 76
Lake Tanganyika (Africa), 76–78
Lakes, 13, 35, 75–78, 93, 102
Larsen B ice shelf, 37, *38*
Latin America, 91
Lee, Kitack, 21
Lehman, Scott, 43
Lehmann, Johannes, 28
Leroux, Marcel, 30–31
Lindzen, Richard, 62
Loisel, Julie, 46, 48
The Long Thaw (Archer), 59
Loops. *See* Feedback loops
Los Angeles Times (newspaper), 94
Louisiana, 52
Lubrication, 37, 39

M

Ma, Michelle, 90
Mangroves, 49, 105
Marchitto, Thomas, 43
Maritime climates, 13, 41, 105
Massachusetts, 14
McCarthy, Michael, 60
McKitrick, Ross, 44, 56, 63, 66, 84
McWhirter, Sheri, 76
Mekong River (Asia), 93
Meltwater, 38, 46
Merchant/cargo ships, 76
Methane, 4, 51, 94–97, 103, 105
Michaels, Patrick J., 55, 84
Mid-Atlantic coast, 52, 75
Migration, 93

Militias, 69
Mississippi River, 75
Mitigation, 7, 51, 56–57, 103
Models
  on aridity, 70
  features of, 4–5
  of ice sheets, 35–36
  predictions from, 6, 28, 63, 70–71
  of sea level rise, 43–44, 46
  See also Computer models
Monsoons, 16, 60, 105
Moon, Ban Ki, 69
Moore, Curtis A., 54
Morrison, Grant
  JLA: American Dream, 42
Mote, Philip, 34
Mouland, Bill, 33
Moulins, 38, 39, 105
Mount Kilimanjaro (Tanzania), 34, 35
Mount Pinatubo (Philippines), 3, 5
Mountain glacial melt, 34–35, *35*
Murphy, Kim, 94
Murray, Senan, 77

N

NASA Goddard Institute for Space Studies, 5
NASA Jet Propulsion Laboratory, 31
NASA (National Aeronautics and Space Administration), 77
National Center for Atmospheric Research (NCAR), 74
National Geographic (magazine), 60
National glacier inventory, 33
National Resources Defense Council (NRDC), 87
Natural factors, 3, 4, 5
Natural gas, 4
Naylor, Rosamond, 82
NCAR (National Center for Atmospheric Research), 74
Negative feedback loops, 24, 62
Nepal, 59
Netherlands, *52,* 53, 101
New ice age, 40, 41
New Orleans (LA), 63, *64–65*
New York (NY), 49, 51, 75
New York (state), 14

Nickels, Greg, 78
Niger, 77
Niger River (West Africa), 74
Nigeria, 54, 56, 77, 101
North Atlantic current, 40, 41, 42, 43, 106
North Carolina, 51
Northern Hemisphere, 41, 78
NRDC (National Resources Defense Council), 87
Nuclear power, 89
Nurek Dam (Tajikistan), 95
Nutrients, 49, 75, 78

O

Obama, Barack, 61
Oceans
    carbon absorption and, 20–22
    changes in level, 30, 31, 32–34
    climate and, 13–14, 99–100
    cooling, 31
    heat absorption and, 19, 20, 31–32
    rivers and, 75
    sea ice extent, 25*t*
    warming, 31, 36*t*, 44
    wave power from, 90–91
    wetlands and, 49
Oil, 4
Oil, Water, and Climate (Gautier), 63, 78, 92
Okeowo, Alexis, 60
Olympics, 61
O'Reilly, Catherine, 77
Ortiz, Joseph, 43
Overgrazing, 77
Overpeck, Jonathan, 43, 44, 67
Ozone, 7

P

Pacific Coast, 50*t*, 90
Pacific Northwest, 16, 79, 91
Pacific Ocean, 4, 54, 55, 56, 66, 74
Pakistan, 95
Pani River (Asia), 95
Peak stream discharge, 14
Peat moss, 26, 46, 47
Permafrost, 26
Petroleum, 87
Philadelphia (PA), 49, 75
Philippines, 3, 5
Plant waste, 21
Plants, 11, 62, 94
Plate tectonics, 3, 56
Polar bears, 6, 33
Polar ice melting, *27*
Poles, 25, 66
Pollution, 87
Ponds, 37
Popular culture, 42
Porter, Howard, 42
Powell, James Lawrence
    Dead Pool, 79
Power plants, 92
Precipitation
    climate change and, 74–75
    decreases in, 70–71
    in feedback loops, 23, 62
    in heat cycle, 20
    increases in, 59, 61
    lake-effect snow, 12–13
    projected changes, 68*t*
    temperature and, 79
    See also Rainfall
Prendergast, Alan, 62
Procrastination penalty, 6, 8

Q

Quality of life, 15

R

Rahmstorf, Stefan, 46
Rain forests, 71
Rainfall
    global warming and, 62
    increases in, 6, 59, 60, 102
    predictions regarding, 67
    temperatures and, 70, 71
    See also Precipitation
Relative humidity, 23
Renewables Global Status Report 2006 Update, 91
Reservoirs, 82, 91, 95
Resources, 69
Retallack, Greg, 16
Revkin, Andrew C., 39
River mouths discharge, 80*t*

Rivers, 74–75, 80*t*, 102
Roach, John, 41
Rocky Mountains, 79
Rogun Dam (Tajikistan), 95
Rome (ancient), 87
Russell, Randy, 24

S

Salinity, 54
Salt, 54
Salt marshes, 49
Saltwater, 49, 75
Savannahs, 71, 102
Scavia, Don, 76
Scientists, 4–6, 44
Scotland, 90
Sea ice, 25*t*, 32
Sea-level rise
   on Atlantic Coast, 47*t*
   contributors to, 36*t*
   dikes and, *52*
   global, 53–54
   impacts of, 6–7
   for islands, 54, 55
   predictions regarding, 42–48
   thermal expansion and, 30–32, 101
   on U.S. coastlines, 47*t*, 49–53
Sea of Japan, 22
Seager, Richard, 41, 70
Seattle Times (newspaper), 90
Seattle (WA), 16, 41, 79
Seismic activity, 94
Severn estuary (Wales), 89
Shipping, 76
Siberia, 14
Simms, Andrew, 55
Skeptical Science (Web site), 23
Snow, 3, 10, 24, 26, 61
Snowdrifts, 33
Snowmelt, 78, 79, 81
Snowpack, 78–79, 81, 106
Soil, 28
Soil salinity, 54
Solar radiation, 3
Southwest, 62, 70, 81, 102
Spawning, 93
Specific heat, 13, 19, 99, 106
Statue of Liberty, 42
Stewart, Robert, 19, 20
Storms
   rising seas and, 49
   surges from, 51, 54
   temperature rise and, 66, 67
   tropical, 17
   wetlands and, 51
Stormwater runoff, 14
Strand, Stuart, 21
Stratosphere, 3, 7
Sublimation, 11, 106
Sudan, 69
Summer sea ice, 24, 26
Summers, 14
The Sun (newspaper), 42
Sunlight, 3, 4, 11, 20, 23, 24
Surges, 51, 54
Swimming, *83*

T

Tajikistan, 95
Taken by Storm (Essex), 44
Tambopata (Peru), 71
Tanzania, 34, 35
Tarawa (Pacific island), 54
Tasmania, 44
Taylor, Richard, 82
Tebua (Pacific island), 54
Temperature
   changes in, 5, 17, 23, 76
   droughts and, 70, 71
   extremes, 14
   impact on currents, 41
   moderation of, 41
   snowpack and, 79
   water and, 10, 13, 70
Thermal expansion, 30, 31, 32, 44, 101, 106
Thermal power plants, 92
Thompson, Lonnie, 34
Tidal power, 88–90, 102, 106
Tidal turbines, 88, *89*
Tides, 55
Tokyo (Japan), 42
Transpiration, 11, 106
Trees, 26
Tropics, 82
Tucurui (Brazil), 96

Turbines, 88, *89*, 95
Turner, Eugene, 51
Tuvalu (Pacific atolls), 55

U

United Kingdom (UK), 97
United Nations, 69
United States
    climate change in, 85
    droughts and, 70, 75, 79
    electrical production in, 92*t*
    hydroelectric power in, 91
    sea-level rise and, 49–53, 101
Urbanization, 14, 15
U.S. Climate Change Program, 45–46
U.S. coastlines, 49–53
U.S. East Coast, 7
U.S. Midwest, 75, 76
U.S. News & World Report (magazine), 85
U.S. Southwest, 62, 70, 81, 102
Uzbekistan, 95

V

Vakhsh River (Asia), 95
Van de Wal, Roderik S.W., 38, 39
Vancouver (B.C.), 61
Vapor. *See* Water vapor
Volcanoes, 3, 5, 7

W

Wall Street Journal (newspaper), 94
Walruses, 6
Warming
    El Niño and, 15
    human factors causing, 5–6
    mitigation of, 7–8
    natural factors causing, 3–4
    of oceans, 31, 32
    from water vapor, 23
    See also Global warming
Washington (D.C.), 61
Washington Post (newspaper), 69
Water
    changes in use of, 80*t*
    climate and, 10, 99–101
    hydrologic cycle and, 9, 10
    locations/sources, 12
    specific heat of, 13
    temperature changes, 13, 17
Water control networks, 53
Water power, 87
    See also Hydroelectric power; Hydropower; Tidal power; Wave power
Water supply
    contamination, 55, 75
    for crops, 82–85
    groundwater and, 81–82
    lakes and, 75–78
    melting snow and, 78–81
    rivers and, 74–75
Water vapor, 10–11, 20, 23–24, 28, 63, 99, 100
Water wheels, 87–88
Wave power, 90–91, 102, 106
Weather, 4, 61
Weaver, Andrew, 33–34, 43
    Keeping Our Cool, 42, 61
Web sites
    Earth Observatory, 11
    Skeptical Science, 23
    WW2010, 12
Webster, Peter, 17
Wentz, Frank, 59, 62
West Africa, 54, 60, 67, 77
Wetlands, 49, 51, 93, 106
Wildlife, 6
Willis, Josh, 31
Wilnis (Netherlands), *52*
Wind, 4, 14, 20
Wind power technology, 88, 91
Wind turbines, 88
World Wildlife Fund-Australia, 71
WW2010 (Web site), 12

Y

Yangtze River (China), 75
Yellow River (China), 74
Yin, Jianjun, 51, 52
Yosemite National Park, 35

Z

Zhang, Xuebin, 60
Zipingpu Dam, China, 93
Zwiers, Francis, 60

About the Author

Noah Berlatsky has edited books for Greenhaven Press's Global Viewpoints, Opposing Viewpoints, and At Issue series. His writing and criticism appear in *The Chicago Reader, Reason, The Comics Journal, The Knoxville Metropulse,* Comixology.com, *Splice Today,* and other publications. He blogs at The Hooded Utilitarian.

DISCARD

APR 2 7 2012 ✓

GC 89 .W38 2011
Berlatsky, Noah.
Water and ice
1/2012